# FLORA OF TROPICAL EA

T0251630

## LORANTHACEAE

R.M. Polhill & D. Wiens*

Shrubs hemiparasitic on the branches (or elsewhere rarely the roots) of other dicotyledons, attached by woody haustoria, with or without surface runners producing secondary haustoria, generally evergreen. Leaves opposite, alternate or whorled, simple, entire, often leathery or rather fleshy, estipulate. Flowers bisexual (rarely unisexual elsewhere), variously borne singly or in racemes, umbels or heads (elsewhere often in 3-flowered dichasia) in axils, at old nodes or terminally, in Africa often large and brightly coloured; bract usually 1 per flower, cupular or unilateral, with a small to leafy limb. Calyx rim-like to tubular, entire to shortly toothed. Petals free or united, 4–5 in Africa, valvate, radially symmetrical or opening with a unilateral split. Stamens as many as petals, epipetalous; anthers basifixed in Africa, 2–4-thecous, sometimes locellate, opening lengthwise. Ovary (receptacle) inferior, unilocular or in more primitive genera with several obscure locules; ovules not differentiated; embryo sacs formed at the base of ovary (mamelon); style and stigma simple. Fruit a berry in Africa (rarely dry and winged elsewhere). Seeds without a testa, normally surrounded by a sticky layer developed from the fruit-wall.

77 genera and about 950 species, widely distributed from the tropics to temperate regions, particularly in the south.

The aerial stem-parasites commonly known as mistletoes belong to several families, of which the major ones are Loranthaceae and Viscaceae. These two families were formerly regarded as subfamilies, but are now considered to have originated separately, the Loranthaceae related to Olacaceae, the Viscaceae to Santalaceae. They differ principally in that Loranthaceae have hermaphrodite flowers with a calyx and showy corolla, while Viscaceae have small inconspicuous flowers and only one perianth whorl. The precarious mode of establishment is, however, similar in most genera of both. The sticky seeds are dispersed by birds, in Africa principally tinkerbirds. The modified hypocotyl forms a pad that adheres to the host-branch to form the haustorium. This penetrates the host to connect with its sap stream to permit further growth of the seedling; secondary haustoria are sometimes produced on runners spreading over the host-branches, as seen in *Plicosepalus* and *Vanwykia*.

The least specialised genera of Loranthaceae occur in southern S. America, New Zealand and Australia. Two genera, *Helixanthera* and *Taxillus*, occur both in Asia and Africa, but the other 21 genera now recognised in Africa, Arabia and Madagascar are all restricted to the region and seem to have undergone most of their radiation there. They may be attributed to two groups of genera. The Tapinanthoid group (genera 1–10) has simple or irregularly branched hairs. The Taxilloid group (genera 11–16) has hairs with whorls of branches (stellate or dendritic). The bracts of these two groups also tend to differ, cupular in the first, unilateral in the latter. The currently accepted genera are largely coincident with the sections adopted in much of the African literature, notably the masterly account by Sprague in the Flora of Tropical Africa (1910–1911), modified from the earlier works of Engler, who described the majority of the common species from tropical Africa around the turn of the century. Segregate classifications have been advocated by van Tieghem in the 1890s, Danser in the 1930s, Balle in the 1950s–1960s and by ourselves and our co-workers since the 1970s.

* Department of Biological Sciences, University of Utah, Salt Lake City, Utah, USA.

The distinction of the genera is based nowadays principally on modifications of the flowers related to their mode of pollination. Flowers of the less specialised genera in Africa, notably *Helixanthera*, *Taxillus* and *Vanwykia* in the Flora area, open spontaneously or with little probing. In more advanced genera of both the Tapinanthoid and Taxilloid groups the flowers are increasingly complicated with special mechanisms that can only be manipulated by birds and various signals are developed to attract their attention. In most cases vents appear in the mature bud just below where the anthers are held in the tip of the bud, with the stamens and corolla-lobes held under tension. The region is often indicated by contrasting bands of colour on the corolla and on parts visible through the vents. Probing by the beak of the bird causes the corolla to split open explosively, either radially or unilaterally with a distinct V-split, dusting the bird's head with pollen. Further refinements occur with a more economic and precise directing of the pollen puff. These have occurred with a remarkable degree of convergent evolution in the two groups of genera, such that species of *Agelanthus* and *Phragmanthera* have been included within a broad concept of *Tapinanthus* until recently.

The pollination strategy in *Globimetula* is quite different. Here the distinct head of the mature bud darkens and when pecked the petals coil outwards to reveal vents at the base of the staminal column, probing of which causes the stamens to coil inwards explosively, bending the style in the same direction. The species of *Tapinanthus* (in the strict sense used here) also have a bud-head that darkens at maturity. Pecking causes a V-shaped rent and precise deposition of pollen and inflexion of the style. The divergence of ecogeographic races and species is often marked, among other features, by shifts in signals given to visiting birds. It is noticeable that the modifications in flowers with swollen bud-heads tend to occur on that structure, with various ridges, wings and crowns, whereas those with vents show more variation related to colour-banding, shape and internal hardening of the corolla-lobes.

For further discussion of the biology and biogeography see Kuijt, The Biology of Parasitic Flowering Plants (1969); Barlow & Wiens, The cytogeography of loranthaceous mistletoes, in Taxon 20: 291–312 (1971); Visser, South African Parasitic Flowering Plants (1981); Calder & Bernhardt (eds), The Biology of Mistletoes (1983); Feehan, Explosive flower opening in ornithophily: a study of pollination mechanisms in some Central African Loranthaceae, in J.L.S. 90: 129–144 (1985); Polhill, Speciation patterns in African Loranthaceae, in Holm-Nielsen, Nielsen & Balslev (eds), Tropical Forests: 221–236 (1989); Polhill & Wiens, Mistletoes Afr. (1998).

1. Hairs simple, irregularly branched or lacking · · · · · · · · · · · · · · · · · · · · · · 2
   Hairs stellate or dendritic with whorls of branches, rarely
       present only on youngest parts, sometimes with
       subsimple hairs admixed · · · · · · · · · · · · · · · · · · · · · · · · · · · · · · · · 11
2. Flowers in racemes or spikes, sometimes crowded at apex
       of peduncle; petals 4, separate, coherent below, radially
       spreading to reflexed from the middle where
       the short erect stamens are generally attached · · · · · ·      1. **Helixanthera**
   Flowers borne in heads, umbels, clusters or singly; petals
       not as above, usually joined into a tube below or if free
       (*Plicosepalus* in part) then 5 and strongly ridged inside · · · · · · · · · · · · ·3*
3. Lower part of petals and corolla-tube if present with series
       of strong paired folds inside; secondary haustoria
       present on extensive surface runners · · · · · · · · · · · ·      2. **Plicosepalus**
   Petals without marked folds inside; plants with a single
       haustorial attachment · · · · · · · · · · · · · · · · · · · · · · · · · · · · · · · · · · 4
4. Corolla 6–9 cm. long, when open with 4 radially arranged
       reflexing lobes, the long stamens remaining weakly
       attached to the style by connective-appendages · · · · ·      3. **Emelianthe**
   Corolla smaller, opening with an explosive release of
       pollen, the short filaments inflexed or inrolled · · · · · · · · · · · · · · · · · · 5

* *Oedina pendens* has dendritic hairs only on the youngest parts and might key here, but has racemes with large long-tubed red flowers.

5. Corolla-lobes inrolled at anthesis ·················    6. **Oliverella**
    Corolla-lobes erect, reflexed or rolled outwards at
    anthesis ············································· 6
6. Corolla-lobes 4 ·······························    7. **Englerina**
    Corolla-lobes 5 ······································· 7
7. Corolla usually banded with 1–several contrasting
    colours, with vents opening below the corolla-tip in
    mature buds, the lobes generally remaining erect,
    sometimes recurved or loosely looped ················· 8
    Corolla reddish, sometimes white or greenish towards
    extremities or spotted, with tips of buds swollen and
    darkening at maturity, without vents below, opening
    explosively with lobes generally abruptly reflexed about
    the middle or tightly rolled outwards (elsewhere
    occasionally erect) ································· 10
8. Corolla-lobes much shorter than the tube ···········    8. **Agelanthus**
    Corolla-lobes longer than the tube ······················· 9
9. Corolla-buds straight, with a distinct tube developing a V-
    split (the other lobes sometimes separating to this point
    in dried specimens) ·························    4. **Oncocalyx**
    Corolla-buds bent at base (not always in *S. curta* from NE.
    Tanzania), with a very short tube split to base without a
    distinct V ·······························    5. **Spragueanella**
10. Corolla-lobes reflexed in E. Africa; stigma small, ovoid to
    globose ·································    9. **Tapinanthus**
    Corolla-lobes coiled outwards at anthesis; stigma top-
    shaped to peltate ·····················    10. **Globimetula**
11. Flowers in extended spikes or racemes ························· 12
    Flowers in umbels, clusters or heads ························· 13
12. Corolla yellow to red without dark markings; tip of
    corolla-bud narrow, enclosing elongate anthers;
    montane ·······························    13. **Oedina**
    Corolla orange to pink or red with dark glandular spots at
    base of lobes; tip of corolla-bud obovoid, enclosing
    small anthers with unequal thecae (more elongate only
    in *O. schliebeniana*) ·······················    14. **Oncella**
13. Corolla opening radially, the lobes as long as to distinctly
    longer than the tube; filaments inserted well up the
    corolla-lobes, elongate, the upper part thickened,
    generally articulated and breaking off ············    15. **Erianthemum**
    Corolla with a short V-slit at anthesis, the lobes shorter
    than the tube; filaments attached ± at the base of the
    lobes, short, isodiametric, erect or inrolled ··················· 14
14. Anthers locellate, the thecae subdivided into a series of
    small chambers; stems from a single haustorial
    attachment ·······························    16. **Phragmanthera**
    Anthers not chambered (in Africa); secondary
    haustoria developed on extensive surface runners ··············· 15
15. Corolla slightly curved in bud, slightly inflated medially,
    shortly hairy; stamens separate; style glabrous ······    11. **Taxillus**
    Corolla straight-sided, tomentose; filaments short,
    incurved, the anthers forming a central pollen-mass;
    style hairy on lower half ·····················    12. **Vanwykia**

## 1. HELIXANTHERA

Lour., Fl. Cochinch.: 142 (1790); Polh. & Wiens, Mistletoes Afr.: 79 (1998)

*Loranthus* L. sect. *Acrostachys* Benth. & Hook.f., G.P. 3: 208 (1880)
*Sycophila* Tiegh. in Bull. Soc. Bot. Fr. 41: 485 (1894)
*Acrostachys* (Benth. & Hook.f.) Tiegh. in Bull. Soc. Bot. Fr. 41: 504 (1894)

Mostly small shrubs from a single haustorial attachment in Africa, essentially glabrous in Africa, elsewhere sometimes with a scurfy indumentum including scales, short stiff hairs and small branched hairs; twigs terete, flattened or angular. Leaves opposite to alternate, sessile to petiolate, usually penninerved. Flowers usually in terminal and/or axillary racemes or spikes, sometimes crowded at the apex of peduncle, subumbellate, 4(–7 in Asia)-merous; bract unilateral to cupular, with a small ovate-triangular limb, often umbonate. Calyx rim-like to cupular, subentire to slightly toothed. Petals separate, white, yellow or red, but not conspicuously banded; lower part erect, generally thickened, sometimes callused inside at top; upper longer part caudate to spathulate, spreading to reflexed. Stamens erect, normally arising from top of lower part of petal, sometimes higher; anthers shorter or longer than filament, 2–4-thecous, sometimes locellate. Style angular, slender with a small stigma in Africa. Berries round to ellipsoid, smooth or warted.

About 45 species, 12 in tropical Africa, the rest in tropical Asia, south-east as far as Sulawesi.

1. Anthers locellate (fig. 1/4); racemes terminal and axillary;
    petiole glabrous; forests · · · · · · · · · · · · · · · · · · · · · · · · · · · · · · · · · · 2
   Anthers not chambered; racemes terminal; petiole (if well
    developed) generally glandular-hairy in the groove; open
    country · · · · · · · · · · · · · · · · · · · · · · · · · · · · · · · · · · · · · · · · · 3
2. Petals white or cream, ± tinged pink; receptacle and berries
    usually smooth · · · · · · · · · · · · · · · · · · · · · · · · · · · · ·    1. *H. mannii*
   Petals yellow to orange-yellow; receptacle and young fruits
    covered in small warts · · · · · · · · · · · · · · · · · · · · · · · · ·    2. *H. verruculosa*
3. Leaves mostly alternate on long shoots, crowded on short
    shoots; racemes secund · · · · · · · · · · · · · · · · · · · · · · · ·    5. *H. thomsonii*
   Leaves mostly opposite or subopposite, the pairs well spaced
    except sometimes just below the inflorescence; flowers
    spirally disposed · · · · · · · · · · · · · · · · · · · · · · · · · · · · · · · · · · · · · 4
4. Flowers 4–9 mm. long · · · · · · · · · · · · · · · · · · · · · · · · · ·    3. *H. kirkii*
   Flowers 20–23 mm. long · · · · · · · · · · · · · · · · · · · · · · · ·    4. *H. tetrapartita*

1. **H. mannii** (*Oliv.*) *Danser* in Verh. K. Akad. Wet., sect. 2, 29(6): 58 (1933); Balle in F.W.T.A., ed. 2, 1: 659 (1958) & in Bol. Soc. Brot., sér. 2, 38: 42, t. 3/1–2, 4/1 (1964) & in Fl. Cameroun 23: 23, t. 5/1–8 (1982); Polh. & Wiens, Mistletoes Afr.: 81, photo. 16, fig. 2A (1998). Type: São Tomé, *Mann* 1075 (K!, holo., P!, iso.)

Stems spreading and pendent, up to 1 m. or so; branchlets 1–3 mm. diameter, reddish to purple-brown. Leaves opposite, subopposite or rarely ternate; petiole 4–12 mm. long; lamina chartaceous, narrowly elliptic-oblong to elliptic, ovate or oblong-oblanceolate, 6–10(–14) cm. long, 1–4 cm. wide, usually pointed, often slightly acuminate at the apex, with 6–10(–12) pairs of fine lateral nerves. Racemes terminal and in many axils, 2–10(–15) cm. long, 20–50-flowered; pedicel 1–5 mm. long; bract ovate-triangular, sometimes slightly gibbous, 0.7–1.5 mm. long, glandular-ciliolate. Receptacle 1–3 mm. long; calyx entire to slightly toothed, 0.3–1 mm. long, glandular-ciliolate. Corolla in bud 4-angular in lower part, ± shouldered to the tapered upper part; petals white to pinkish, linear, (4–)8–18(–24) mm. long, thickened, grooved and sometimes papillate below filament-insertion. Stamen-filament arising from behind a flap-like petal-appendage; anthers locellate, 5–9 mm. long. Style 3–7 mm. long. Berry tinged red, ellipsoid, 5–6 mm. long, generally smooth, but in Gabon often with fine lines of longitudinal pimples. Fig. 1/1–5 (p. 6).

UGANDA. Toro District: Bwamba, Oct. 1925, *Fyffe* 21!
DISTR. U 2; westwards to Nigeria, São Tomé and Angola
HAB. Rain-forest, on a wide variety of hosts; ± 800 m.

SYN. *Loranthus mannii* Oliv. in J.L.S. 7: 101 (1863) & in Hook., Ic. Pl. 14, t. 1303 (1880); Engl.
    in E.J. 20: 81 (1894); Sprague in F.T.A. 6(1): 274 (1910); Balle in F.C.B. 1: 316, t. 33
    (1948)
  *L. mannii* Oliv. var. *combretoideus* Engl. in E.J. 20: 81 (1894). Type: Angola, Pungo Andongo,
    Calemba Lake, *Welwitsch* 4852 (COI!, holo., B!, BM!, K!, P!, iso.)
  *Sycophila combretoides* Tiegh. in Bull. Soc. Bot. Fr. 41: 486 (1894). Type as for *Helixanthera
    mannii* var. *combretoideus*
  *S. mannii* (Oliv.) Tiegh. in Bull. Soc. Bot. Fr. 41: 487 (1894)
  *Loranthus combretoides* (Tiegh.) Engl. in E. & P. Pf., Nachtr. 1: 128 (1897); Sprague in F.T.A.
    6(1): 274 (1910)
  *Helixanthera combretoides* (Tiegh.) Danser in Verh. K. Akad. Wet., sect. 2, 29(6): 56 (1933)

2. **H. verruculosa** *Wiens & Polh.*, Mistletoes Afr.: 83 (1998). Type: Tanzania, Iringa District, Mufindi, Livalonge, *Congdon* 28 (K!, holo., BR!, DSM!, EA!, iso.)

Stems well branched, up to 80 cm. long; branchlets 1–3 mm. diameter, brown to purple-brown. Leaves opposite or subopposite; petiole 5–10 mm. long; lamina thinly textured, dark green at maturity, elliptic to elliptic-oblong, 5–10 cm. long, 2–5 cm. wide, slightly acuminate at the apex, cuneate to rounded at the base, with 6–10 pairs of lateral nerves. Racemes terminal and axillary, 4–8 cm. long, mostly 30–50-flowered; pedicel 2–3 mm. long; bract ovate-triangular, 1 mm. long, umbonate, ciliolate. Receptacle narrowly obconic, 1.5 mm. long, covered in small warts; calyx 0.5–0.8 mm. long, ciliolate. Corolla in bud 4-angular in lower quarter, slightly shouldered to tapering upper part; petals yellow to orange-yellow, linear, 6–7 mm. long, thickened and papillate below filament-insertion, reflexed above. Stamen-filament arising from nectariferous callus, 2–3 mm. long; anthers locellate, 1.5–3 mm. long. Style 5–7 mm. long. Berry obovoid-ellipsoid, 5–6 mm. long, 3.5–4 mm. in diameter, verruculose, becoming smooth and red at maturity; seed greenish-yellow, translucent at the base.

TANZANIA. Iringa District: Kigogo Forest Reserve, Dec. 1981, *Rodgers & Hall* 2827! & Kibao, 5
    Jan. 1986, *Wiens et al.* 6543! & Livalonge Estate, 4 May 1986, *Polhill et al.* 5263!
DISTR. T 7; not known elsewhere
HAB. Upland rain-forest, on various hosts, but commonly *Maesa*; 1450–1900 m.

3. **H. kirkii** (*Oliv.*) *Danser* in Verh. K. Akad. Wet., sect. 2, 29(6): 57 (1933); Polh. & Wiens in U.K.W.F., ed. 2: 154 (1994) & Mistletoes Afr.: 83, photo. 17, fig. 2B (1998). Type: Tanzania, Mikindani District, Ruvuma Bay, *Kirk* (K!, holo., P!, fragment)

Stems laxly branched, to 60 cm. or so; branchlets thickening to 3–5 mm. diameter, greenish turning reddish brown to grey. Leaves mostly opposite or subopposite, sometimes one alternating with pairs; petiole (0.5–)0.8–2.5 cm. long, glandular-puberulous in the groove; lamina slightly fleshy, usually ovate to elliptic, sometimes lanceolate or subcircular, 3–10 cm. long, 1.5–7 cm. wide, blunt or pointed at the apex, with ± 6–12 pairs of lateral nerves. Racemes terminal, 5–10(–15) cm. long, mostly subdensely 40–100-flowered, sometimes with scurfy indumentum; pedicel 3–5(–8 in fruit) mm. long; bract broadly ovate, 1–1.5 mm. long. Receptacle 1–2 mm. long; calyx subentire. Corolla in bud narrowed one-third to halfway up, slightly quadrangular below, a little tapered above; petals yellow to orange- or purple-red,

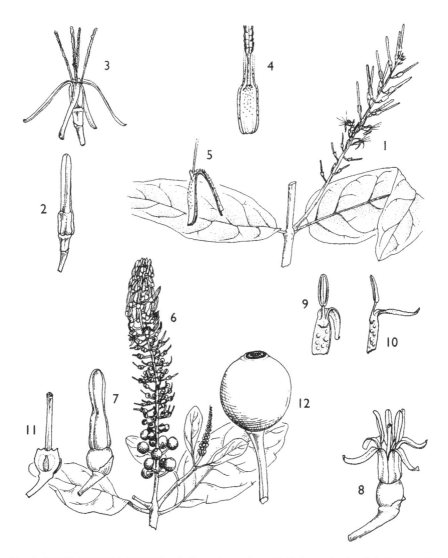

FIG. 1. *HELIXANTHERA MANNII* — **1**, flowering node, × ²/₃; **2**, flower bud, × 2; **3**, flower, × 2; **4, 5**, base of petal showing stamen insertion, front and side views, × 4. *HELIXANTHERA KIRKII* — **6**, habit, × ²/₃; **7**, flower bud, × 4; **8**, flower, × 4; **9, 10**, petal and stamen, front and side views, × 4; **11**, section of gynoecium, × 4; **12**, fruit, × 4. 1–5, from *Sanford* 6262; 6, from *Drummond & Hemsley* 2316; 7–12, from *Faulkner* 1730. Drawn by Christine Grey-Wilson.

sometimes with green, linear-lanceolate, 4–9 mm. long, slightly thickened, folded and medially grooved below filament-insertion. Anthers 2–4 mm. long, 1–6 times as long as filament. Style 3–5 mm. long. Berry turning red then black, globose, 5 mm. across. Fig. 1/6–12.

KENYA. Northern Frontier Province: Mt. Kulal, El Kajata gorge, 28 Dec. 1978, *Synnott* 1760!; Masai District: 56 km. on Magadi–Nairobi road, 5 Aug. 1957, *Greenway* 9220!; Kwale District: Waa, 20 Sept. 1982, *Polhill* 4811!

TANZANIA. Mbulu District: Oldeani–Mongala, 16 km. from turn-off on Oldeani coffee farm road, 1 Apr. 1964, *Verdcourt* 4020!; Lushoto District: Mkomazi R. on Mkomazi–Buiko road, 30

Apr. 1953, *Drummond & Hemsley* 2316!; Uzaramo District: Dar es Salaam, University, 3 June 1966, *Mwasumbi* 10141!

DISTR. **K** 1, 4, 6, 7; **T** 2, 3, 5, 6, 8; along the coast from S. Somalia to N. Mozambique, inland to the Great Rift Valley of Kenya and Tanzania, then disjunctly to S. Angola

HAB. Coastal bushland and mixed dry bushland and woodland inland, on various hosts, commonly *Cordia*, *Grewia* and *Markhamia*; 0–1350 m.

SYN. *Loranthus kirkii* Oliv. in J.L.S. 7: 101 (1863) & in Hook., Ic. Pl. 14, t. 1309 (1880); Engl. in E.J. 20: 129 (1894) & P.O.A. C: 167 (1895); Sprague in F.T.A. C(1): 276 (1910); F.D.O.-A. 2: 160 (1932); I.T.U.L.: 280 (1949); U.K.W.F.: 330 (1974)

    *L. kirkii* Oliv. var. *ciliatus* Engl. in E.J. 20: 129 (1894) & P.O.A. C: 167 (1895); F.D.O.-A. 2: 161 (1932). Type: Kenya/Tanzania, without locality, *Fischer* I. 302 (HBG!, holo., B!, iso.)

    *L. kirkii* Oliv. var. *populifolius* Engl. in E.J. 20: 130 (1894) & P.O.A. C: 167 (1895); F.D.O.-A. 2: 161 (1932). Type: Tanzania, Bagamoyo District, Kingani [Kigani] R., *Hildebrandt* 1141 (B!, holo.)

    *Acrostachys kirkii* (Oliv.) Tiegh. in Bull. Soc. Bot. Fr. 41: 504 (1894)

    *Loranthus stefaninii* (fig.) & *somalensis* (descr.) Chiov., Result. Sci. Miss. Stef.-Paoli, Coll. Bot. 1: 155, t. 15B (1916); E.P.A.: 32 (1953). Type: Somalia, between Torda and Margherita, *Paoli* 342 (FT, holo.)

    *Helixanthera somalensis* (Chiov.) Danser in Verh. K. Akad. Wet., sect. 2, 29(6): 59 (1933)

NOTE. The E. African coastal populations have dull reddish flowers, those from inland are bright yellow or orange and those from Somalia are recorded as yellow and green. The Angolan populations have distinctly more pointed leaves than those from E. Africa, and, as in E. Africa, there may be a rather similar variation in flower colour, the coastal plants tending to have redder flowers at least at maturity.

4. **H. tetrapartita** (*E.A. Bruce*) *Wiens & Polh.* in Lebrun & Stork, Énum. Pl. Fl. Afr. Trop. 2: 169 (1992) & Mistletoes Afr.: 85 (1998). Type: Tanzania, Dodoma District, Dodoma–Iringa, near Great Ruaha R., *St. Clair-Thompson* 369 (K!, holo.)

Stems to 50 cm. or so; branchlets soon thickening to 3–5 mm. diameter, reddish brown turning grey. Leaves mostly opposite or subopposite, but often a few crowded below inflorescence; petiole 0.5–2 cm. long, glandular-hairy in groove; lamina slightly fleshy, ovate-lanceolate to ovate or broadly elliptic, 3.5–5 cm. long, 2–4 cm. wide, bluntly pointed at the apex, with 3–6 pairs of obscure lateral nerves. Racemes terminal on long and few-leaved short shoots, 7–17 cm. long, laxly 30–50-flowered; pedicel 2–4 mm. long; bract ovate-triangular, slightly gibbous, 1.5 mm. long. Receptacle 2–2.5 mm. long; calyx subentire to slightly toothed, glandular-ciliolate. Corolla in bud quadrangular in lower sixth or seventh, scarcely shouldered to tapered upper part; petals red or yellow, linear, 2–2.3 cm. long, slightly thickened and grooved below filament-insertion. Anthers 8 mm. long, ± as long as filament. Style 14–15 mm. long. Berry red, obovoid, 7 mm. long.

TANZANIA. Mpwapwa, 19 Feb. 1933, *Hornby* 502!; Mbeya District: Ruaha National Park, summit of Magangwe Hill, 12 Mar. 1973, *A. Bjørnstad* 2621!; Njombe District: Lower Ndumbi valley, ± 5 km. from main road, 24 May 1986, *J. & J. Lovett* 756!

DISTR. **T** 5, 7; not known elsewhere

HAB. Deciduous mixed woodland at the interface of *Brachystegia* and *Acacia-Commiphora* associations, recorded on *Commiphora*; 950–1700 m.

SYN. *Loranthus tetrapartitus* E.A. Bruce in K.B. 1935: 278 (1935), as '*tetraparitus*'; T.T.C.L.: 280 (1949)

5. **H. thomsonii** (*Sprague*) *Danser* in Verh. K. Akad. Wet., sect. 2, 29(6): 60 (1933); M.G. Gilbert in Fl. Ethiopia 3: 362, fig. 114.1 (1990); Polh. & Wiens, Mistletoes Afr.: 86, photo. 19 (1998). Type: N. Somalia, Haud, *D. Thomson* 23 (K!, holo.)

Stems to 50 cm. or so; branchlets soon thickening to 4–10 mm. diameter, red-brown to purple, becoming silvery from detaching epidermis and then grey. Leaves mostly alternate on long shoots, crowded on short shoots; petiole 1–10 mm. long, glandular-puberulous in the groove; lamina fleshy, oblanceolate-spathulate, 4–25 mm. long, 1.5–6 mm., subacute to rounded at the apex. Racemes terminal on short shoots, secund, 1–5 cm. long, densely 10–20-flowered; pedicel 0–2 mm. long; bract ovate-triangular, 1–1.5 mm. long. Receptacle 1–2 mm. long; calyx subentire to slightly toothed, glandular-ciliolate. Corolla in bud linear, slightly swollen in lower one-fifth to one-eighth below insertion of the filaments and over anthers near the apex; petals pale greenish turning yellow, very narrowly linear-spathulate, looped near middle, 1.5–3 cm. long, notably thickened only at insertion of filaments. Anthers 2–5 mm. long; filaments very much longer, pink to red, thickened at the base. Style as long as petals. Berry bright red, obovoid, 7 mm. long.

Kenya. Northern Frontier Province: Moyale, 12 Mar. 1952, *Gillett* 12529! & Ramu–Banissa, 29 km. from turning to Banissa, 4 May 1978, *M.G. Gilbert & Thulin* 1436! & Kenya-Somalia border, 23 Apr. 1973, *Oxtoby* in *E.A.* 15338!
Distr. K 1, 4; Ethiopia (Ogaden), Somalia and Yemen
Hab. Deciduous bushland, on various hosts, including *Commiphora* and *Lannea*; 10–1100 m.

Syn. *Loranthus thomsonii* Sprague in F.T.A. 6(1): 276 (1910) & in K.B. 1911: 88 (1911)


## 2. PLICOSEPALUS

Tiegh. in Bull. Soc. Bot. Fr. 41: 504 (1894); Polh. & Wiens, Mistletoes Afr.: 87 (1998)

*Loranthus* L. sect. *Plicopetalus* Benth. & Hook.f., G.P. 3: 208 (1880); Sprague in F.T.A.
     6(1): 258 (1910)
*Loranthus* L. sect. *Tapinostemma* Benth. & Hook.f., G.P. 3: 209 (1880); Sprague in F.T.A.
     6(1): 259 (1910)
*Tapinostemma* (Benth. & Hook.f.) Tiegh. in Bull. Soc. Bot. Fr. 42: 243, 257 (1895)


Stems a few cm. to several m. long, glabrous or rarely with short stiff simple hairs, spreading by haustoria-bearing surface runners. Leaves irregularly opposite to alternate, often clustered on short shoots, sessile to shortly petiolate, leathery to slightly fleshy, palminerved. Flowers (1–)few in umbels in the axils or terminal on short shoots, 5-merous; bract saucer-shaped, with a small umbonate or spurred oblong to ovate-triangular limb, sometimes lacerate. Calyx rim-like to cupular, sometimes split as flower opens. Petals connivent or joined in lower part, slightly to markedly curved in bud, mostly red or yellow, with 2 rows of oblique folds below the filament-insertion, these parts connivent, the upper part narrow below the linear-lanceolate to linear-spathulate tips, reflexing, looped and often twisted; corolla-tube, if present, sometimes with further series of flaps or a lobed ledge. Stamens following curve of style and subequal in length, red at least above; anthers 4-thecous. Style filiform, angular, slightly curved to sinuous, sometimes corrugated near the base, only slightly thickened where anthers appressed in bud; stigma capitate. Berry red or yellow, ellipsoid or urceolate, smooth or warty.

12 species from the Middle East and Arabia, extending down the eastern side of Africa to Angola and South Africa, mostly in dry country, mainly on legumes or Burseraceae.

1. Corolla with a long tube, pinkish tipped green or greenish
     blue; umbels with 1–2 flowers often on very contracted
     stems · · · · · · · · · · · · · · · · · · · · · · · · · · · · · · · · · ·     4. *P. meridianus*
     Corolla of separate petals connivent in lower part, variously
     yellow, red and whitish · · · · · · · · · · · · · · · · · · · · · · · · · · · 2
2. Receptacle (5–)6–8 mm. long; berries 12–15 mm. long · · ·     3. *P. kalachariensis*
     Receptacle 2.5–4(–5) mm. long; berries 9–12 mm. long · · · · · · · · · · · · · 3
3. Leaves all cuneate to a short petiole · · · · · · · · · · · · · · ·     1. *P. curviflorus*
     Leaves cordate to sagittate on long shoots (variably shaped
     and sometimes shortly petiolate on short shoots) · · · · ·     2. *P. sagittifolius*

1. **P. curviflorus** (*Oliv.*) *Tiegh.* in Bull. Soc. Bot. Fr. 41: 504 (1894); Danser in Verh. K. Akad. Wet., sect. 2, 29(6): 100 (1933); Blundell, Wild Fl. E. Afr.: 131, t. 514 (1987), pro parte; M.G. Gilbert in Fl. Ethiopia 3: 364, fig. 114.5 (1990); Polh. & Wiens in U.K.W.F., ed. 2: 154, t. 54 (1994) & Mistletoes Afr.: 88, photo. 8, fig. 3A (1998). Type: Ethiopia, without precise locality, *Plowden* (K!, holo.)

Leaves not markedly dissimilar where clustered on short shoots; petiole 1–3 mm. long; lamina usually linear, oblong-lanceolate or -oblanceolate, atypically obovate, 2–7.5 cm. long, 0.2–1.5(–2) cm. wide, basally cuneate, 3–5-nerved but venation often obscure. Umbels 1–3, axillary, at nodes below or on short shoots, 3–7-flowered; peduncle (1–)3–9 mm. long; pedicels 4–11 mm. long. Receptacle campanulate to urceolate, 2.5–4(–5) mm. long; calyx rim-like, 0.5–1 mm. long. Petals separate, 2.5–3.5 cm. long; basal part below filament-insertion red, yellow or less often white or orange, S-shaped, 5–8 mm. long, with 4–5 paired folds inside; distal part normally red, atypically yellow, recurving, ± twisted, linear-spathulate above. Stamens red; anthers 7–12 mm. long. Berry red, ellipsoid or urceolate, 9–12 mm. long, 7 mm. in diameter, smooth. Fig. 2/1–5 (p. 10).

UGANDA. Acholi District: Chua, above Lututuru, Feb. 1938, *Eggeling* 3494!; Karamoja District: Lokitanyala, Sept. 1963, *Tweedie* 2724!; Mbale District: Kyosoweri [Kyesoweri], 17 Oct. 1955, *Norman* 296!
KENYA. Northern Frontier Province: Dandu, 20 Mar. 1952, *Gillett* 12607!; Kitui District: 50 km. E. of Thika on Kitui road, *Wiens* 4486!; Teita District: Mt. Kasigau, 4 km. S. of Rukanga at eastern base, 2 May 1981, *M.G. & C.I. Gilbert* 6123!
TANZANIA. Maswa District: Moru Kopjes, 31 Dec. 1971, *Greenway & Myles Turner* 13956!; Lushoto District: Mkomazi, 27 May 1962, *Gill* 23!; Iringa District: Ruaha National Park, Nyamakuyu rapids, 6 Aug. 1970, *Thulin & Mhoro* 605!
DISTR. **U** 1, 3; **K** 1–7; **T** 1–3, 7; E. Zaire, Somalia, Ethiopia and along the Red Sea coast to southernmost Egypt, also in the Middle East and the Arabian peninsula
HAB. Deciduous bushland and *Acacia* woodland, almost always on *Acacia*, rarely on *Albizia*; 50–2300 m.

SYN. *Loranthus curviflorus* Oliv. in Hook., Ic. Pl. 14, t. 1304 (1880); P.O.A. C: 167 (1895); Sprague in F.T.A. 6(1): 279 (1910); Engl. in V.E. 3(1): 105, fig. 69A–D (1915); K. Krause in N.B.G.B. 8: 504 (1923); F.D.O.-A. 2: 161 (1932); Engl. & K. Krause in E. & P. Pf., ed. 2, 16B: 148, fig. 70A–D (1935); Balle in F.C.B. 1: 318 (1948); T.T.C.L.: 280 (1949); U.K.W.F.: 330, fig. on 331 (1974), excl. syn. *L. sagittifolius*
  [*L. kalachariensis* sensu T.T.C.L.: 280 (1949), *non* Schinz]

NOTE. Easily recognised in the main northern part of its range, but intergrading to some extent with *P. sagittifolius* in East Africa. The leaves are always cuneate and shortly petiolate at the base. The petals are normally red or yellow at the base and red above, but along the Rift Valley the flowers may be yellow and orange or wholly yellow. The normally narrow, leathery, obscurely veined, often glaucous leaves vary to obovate and nervose near the East African coast, and are also sometimes unusually broad in the Rift Valley. There is a distinct form of the species in the Kenya Highlands above about 1650 m., notably robust, with very leafy floriferous short shoots and generally yellow petals, red at the base or tip; this form, but with more variably coloured flowers, extends over the Serengeti plains of NW. Tanzania.

2. **P. sagittifolius** (*Engl.*) *Danser* in Verh. K. Akad. Wet., sect. 2, 29(6): 100 (1933); M.G. Gilbert in Fl. Ethiopia 3: 364 (1990); Polh. & Wiens in U.K.W.F., ed. 2: 155, t. 55 (1994) & Mistletoes Afr.: 88, photo. 20 (1998). Types: Tanzania, Tanga District, Moa, *Holst* 3103 (B!, syn., COI!, P!, isosyn.) & Lushoto District, Mashewa, *Holst* 8835 (B!, syn., K!, isosyn.)

Leaves heteromorphic, those on long shoots always sessile (except sometimes 1–2 basal pairs) and generally more markedly sagittate or cordate at the base, those on short shoots with a petiole up to 1(–3 inland) mm. long; lamina diversely lanceolate to obovate or circular, 0.8–4.5(–6) cm. long, 0.5–3 cm. wide, rounded to cordate or sagittate at the base, 3–7-nerved, with venation conspicuous to obscure. Umbels

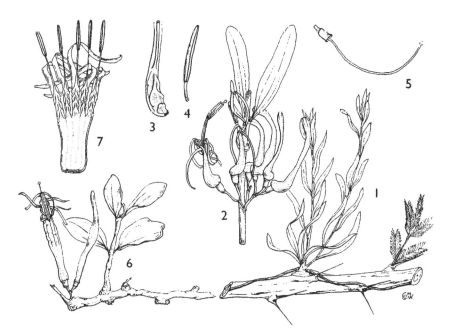

FIG. 2. *PLICOSEPALUS CURVIFLORUS* — **1**, young plant showing surface runners, × ²/₃; **2**, flowering branch, × ²/₃; **3**, base of petal showing stamen insertion, × 2; **4**, anther, × 2; **5**, gynoecium, × 1. *PLICOSEPALUS MERIDIANUS* — **6**, habit, × ²/₃; **7**, corolla opened out, × 1. 1, from *Tweedie* 2489; 2–5, from *Bally & Smith* 14917; 6, 7, from *Tweedie* 2974. Drawn by Christine Grey-Wilson.

normally terminal on leafy short shoots, atypically 1–several at leafy or leafless nodes, 3–7(–12)-flowered; peduncle 5–10(–17) mm. long; pedicels 6–10 mm. long. Receptacle campanulate, 2.5–4 mm. long; calyx rim-like, 0.5–1 mm. long. Petals separate, 3–3.5 cm. long, normally ageing from yellow-green through yellow and orange to red, but inland often yellow and sometimes with a redder basal part; basal part slightly S-shaped, 6–8 mm. long, with 4–7 paired folds inside; distal part recurving, twisted, linear-spathulate above. Stamens orange, pink or red; anthers 7–9 mm. long. Berry red, ellipsoid, 9–10 mm. long, 7 mm. in diameter, smooth.

UGANDA. Acholi District: Rom, Dec. 1935, *Eggeling* 2369!; Karamoja District: Moruangaberu [Emoruangaberru], 14 Jan. 1956, *Dyson-Hudson* 117! & Lodoketemit [Lodoketeminit], 8 Nov. 1962, *Kerfoot* 4465!

KENYA. Northern Frontier Province: 60 km. on Wajir–El Wak road, 29 Apr. 1978, *M.G. Gilbert & Thulin* 1215!; W. Suk District: Kongelai [Kongoli] road, July 1961, *Lucas* 186!; Teita District: 8–11 km. W. of Bura Railway Station, 17 Jan. 1972, *Gillett* 19587!

TANZANIA. Masai District: Kitumbeine Mt., 2 Mar. 1969, *Richards* 24259!; Tanga District: Perani Forest, 12 Aug. 1953, *Drummond & Hemsley* 3707!; Dodoma District: 26 km. on Dodoma–Morogoro road, 12 Apr. 1988, *Bidgood, Mwasumbi & Vollesen* 1030!

DISTR. **U** 1; **K** 1–4, 6, 7; **T** 2, 3, 5–7; coastal regions from S. Somalia to N. Mozambique, inland to the Rift Valley region from S. Ethiopia to S. Tanzania

HAB. Deciduous bushland and *Acacia* woodland, almost invariably on *Acacia*, rarely on *Albizia* or *Commiphora*; 30–2300 m.

SYN. *Loranthus undulatus* Harv. var. *sagittifolius* Engl., P.O.A. C: 167, t. 18A–C (1895), as 'sagittaefolius', & in E.J. 30: 304 (1901)

     *L. sagittifolius* (Engl.) Sprague in F.T.A. 6(1): 278 (1910) & in K.B. 1911: 88 (1911); Engl. in V.E. 3(1): 105, fig. 69J (1915); F.D.O.-A. 2: 161 (1932); Engl. & K. Krause in E. & P. Pf., ed. 2, 16B: 148, fig. 70J (1935); T.T.C.L.: 280 (1949)

NOTE. Agnew, U.K.W.F. (1974), included *P. sagittifolius* in *P. curviflorus* and some herbarium annotations also indicated doubts about the specific distinctness of *P. sagittifolius*. The similarity of populations along the Rift seems to us to result from some secondary introgression. The typical form of *P. sagittifolius* along the E. African coast has very characteristic leafy flowering shoots, the sessile sagittate or cordate primary leaves subtending a short shoot with smaller ± cordate leaves and terminated by an umbel of flowers that turn from yellow to red with age. Inland *P. sagittifolius* can always be separated by the sessile cordate or sagittate leaves on the long shoots, but firstly the leaves on short shoots tend to become more obviously petiolate, the inflorescences tend to be produced severally and later, and the flowers are normally yellow, often with an orange base. On the Rift floor the differences between *P. sagittifolius* and *P. curviflorus* are reduced essentially to the shape of leaves on long shoots, though local populations seem to remain discrete. In an unpublished research paper at Kew, Jacque Triner demonstrated that the anatomy of the leaves was significantly different in the two species, those of *P. curviflorus* distinctly more xeromorphic.

3. **P. kalachariensis** (*Schinz*) *Danser* in Verh. K. Akad. Wet., sect. 2, 29(6): 100 (1933); Wiens & Tölken in F.S.A. 10: 35, fig. 14/2 (1979); Polh. & Wiens, Mistletoes Afr.: 90, photo. 21 (1998). Types: Botswana, Okavango, *Fleck* 307 & Lake Ngami, *Fleck* 314a (Z, syn.)

Leaves light green to glaucous; petiole (0–)1–6 mm. long; lamina variably narrowly oblong to lanceolate, oblanceolate, elliptic-oblong or almost ovate, sometimes slightly curved, 2–12 cm. long, 0.4–5.5 cm. wide, cuneate to rounded or exceptionally somewhat sagittate at the base, 3–7-nerved. Umbels 1–4 in the axils or at the nodes below, 3–6-flowered; peduncle (1–)3–6 mm. long; pedicels 8–12 mm. long. Receptacle obconic, (5–)6–8 mm. long; calyx rim-like, 0.5–1 mm. long. Petals separate, 4–5 cm. long, pink to reddish orange, usually darker towards the base, but sometimes banded or mottled; basal part slightly S-shaped, 9–12 mm. long, with 4–5 paired folds inside; distal part recurving, twisted, linear-oblanceolate to linear-spathulate above. Stamens pinkish; anthers 9–15 mm. long. Berry red, urceolate, 12–15 mm. long, 7–8 mm. in diameter.

KENYA. Kwale District: Gongoni Forest, 3 June 1990, *Luke & S.A. Robertson* 2368!
TANZANIA. Morogoro, Forest Office, 27 Aug. 1956, *Semsei* 2441!; Iringa, just north of the township, 14 July 1956, *Milne-Redhead & Taylor* 11078!; Kilwa District: Kingupira, 7 July 1975, *Vollesen* in *M.R.C.* 2538!
DISTR. **K** 7; **T** 5–8; south to Angola, Namibia and northern parts of South Africa
HAB. Mixed woodland, generally with *Acacia* or *Brachystegia*, usually on legumes, most commonly *Acacia, Albizia, Brachystegia* and *Julbernardia*; 30–1850 m.

SYN. *Loranthus kalachariensis* Schinz in Bull. Herb. Boiss. 4, Append. 3: 53 (1896); Sprague in F.T.A. 6(1): 280 (1910) & Fl. Cap. 5(2): 105 (1915); Engl. in V.E. 3(1): 106, fig. 69E–H (1915); Engl. & K. Krause in E. & P. Pf., ed. 2, 16B: 148, fig. 70E–H (1935)
  *Plicosepalus curviflorus* (Oliv.) Tiegh. var. *kalachariensis* (Schinz) Balle ined.; Vollesen in Opera Bot. 59: 64 (1980), *nom. invalid., comb. non rite publ.*

NOTE. *Plicosepalus kalachariensis* differs from *P. curviflorus* essentially only in its larger flowers and fruits, and if it were not for other taxa in the complex might perhaps be more appropriately ranked as a subspecies. Where the ranges abut in Tanzania there is a slight ecological segregation, with *P. kalachariensis* well established (though not common) in *Brachystegia* woodland, and, like plants from this association further south, the leaves tend to be notably broader than in *P. curviflorus* with 5–7 main pairs of nerves. Nevertheless along the walls of the Rift from Kondoa to Mbeya there are slight indications of a narrow zone of more or less intermediate populations, represented, for example, by Iringa, *Semsei* 2442 & Iringa, Kibebe Farm, *Fison* 151.

4. **P. meridianus** (*Danser*) *Polh. & Wiens* in Nordic Journ. Bot. 5: 222 (1985); Blundell, Wild Fl. E. Afr.: 131, t. 755 (1987); M.G. Gilbert in Fl. Ethiopia 3: 363, fig. 114.2 (1990); Polh. & Wiens in U.K.W.F., ed. 2: 155 (1994) & Mistletoes Afr.: 94, photo. 25, fig. 3F (1998). Type: Tanzania, Dodoma/Iringa District, Njombe R., S. of Kilimatinde between Ilangali [Irangali] and Uwimbi, *von Prittwitz* 175 (B!, holo.)

Leaves opposite to alternate on short stems or directly from the runners or haustoria (appearing as tufts on host); petiole 1–3 mm. long; lamina narrowly elliptic to obovate, 1–4 cm. long, 0.4–2.3 cm. wide, obliquely cuneate at the base, 3–5-nerved from base. Umbels 1–2 from nodes of runners or stems, with or without leaves, or on knobbly contracted stems, 2–4-flowered; peduncle 2–6 mm. long; pedicels 4–10 mm. long. Receptacle obconic-cylindrical, 3–4 mm. long; calyx rim-like, 1–2 mm. long. Corolla 4.5–6.7 cm. long, pinkish tipped green or greenish blue; tube very slightly S-shaped, 2–3 cm. long, narrow, slightly flared in upper oblique part; lower connivent part of the lobes 7–15 mm. long, with normally 6–8 paired folds and papillose inside; lobes distally recurving, linear. Filaments red, at least above; anthers 9–10 mm. long, purple. Berry orange or pink tinged, obovoid, 10 mm. long, 6 mm. in diameter. Fig. 2/6, 7 (p. 10).

UGANDA. Karamoja District: Lokitanyala [Lokitaungyala], Jan. 1955, *Philip* 750! & Moroto R., Feb. 1936, *Eggeling* 2958!
KENYA. W. Suk District: Ortum, Jan. 1965, *Tweedie* 2974!; Machakos District: Mombasa–Nairobi road, km. W. of Kibwezi, 24 Jan. 1972, *Wiens* 4548!; Tana R. District: Garissa–Garsen road, 8.3 km. from turnoff to Bura, 7–9 July 1974, *R.B. & A.J. Faden* 74/1022!
TANZANIA. Musoma District: Seronera National Park, 20 Apr. 1965, *Richards* 20237!; Mbulu District: Katesh, 27 Aug. 1970, *Richards* 25801!; Iringa District: Ruaha valley, 15 km. N. of Mbuyuni on Malolo track, 17 May 1990, *Abdallah & Newton* 2261!
DISTR. U 1; K 1, 2, 4, 7; T 1, 2, 5–7; Ethiopia and Somalia
HAB. Deciduous bushland, generallly on *Commiphora*, sometimes on *Acacia*; 100–1550 m.

SYN. *Tapinostemma meridianum* Danser in N.B.G.B. 11: 216 (1931) & in Verh. K. Akad. Wet., sect. 2, 29(6): 122 (1933)
    *Loranthus ngurukizi* E.A. Bruce in K.B. 1933: 470 (1933). Type: Tanzania, Dodoma District, Manyoni, *B.D. Burtt* 3419 (K!, holo.)
    *L. meridianus* (Danser) K. Krause in E. & P. Pf., ed. 2, 16B: 148 (1935); E.A. Bruce in K.B. 1936: 475 (1936); T.T.C.L.: 283 (1949); U.K.W.F.: 332 (1974)
    [*L. acaciae* sensu U.K.W.F.: 330 (1974), *non* Zucc.]

NOTE. The runners have been likened to the roots of epiphytic orchids, and the tendency to produce tufts of leaves and flowers on such short stems that they appear to emerge from the host is curious.

## 3. EMELIANTHE

Danser in Verh. K. Akad. Wet., sect. 2, 29(6): 53 (1933), pro parte excl. *E. galpinii*; Polh. & Wiens, Mistletoes Afr.: 96 (1998)

*Loranthus* L. sect. *Tetrameri* Sprague in F.T.A. 6(1): 264 (1910)

Shrubs 0.5–2 m., with a single haustorial attachment, glabrous. Leaves mostly alternate and also crowded on short shoots, penninerved. Flowers 2–4, umbellate on a short peduncle or clustered on short shoots of generally leafless branches, shortly pedicellate; bract ovate-triangular from a saucer-shaped based. Calyx cupular. Corolla joined about halfway, 4-lobed, radially symmetrical, with fluted vents opening at top of tube, not explosive when opened, the lobes recurving. Stamens inserted above the base of the corolla-lobes, remaining weakly attached to the style-tip by connective-appendage; anthers 4-thecous, with a conspicuous bifid connective-appendage. Style filiform; stigma capitate. Berry obovoid.

A single species in the drier parts of E. and NE. Africa.

The flowers are more regular than *Plicosepalus*, with vents opening on all sides, the tips recurving to reveal the bright red filaments inside. At anthesis the connective-appendages remain adherent to the tip of the style, so that the anthers and stigma remain in proximity. The pollen is not released explosively.

**E. panganensis** (*Engl.*) *Danser* in Verh. K. Akad. Wet., sect. 2, 29(6): 53 (1933); Blundell, Wild Fl. E. Afr.: 129, t. 511 (1987); M.G. Gilbert in Fl. Ethiopia 3: 365, fig. 114.6 (1990); Polh. & Wiens in U.K.W.F., ed. 2: 155, t. 55 (1994) & Mistletoes Afr.: 96 (1998). Lectotype, chosen by Polh. & Wiens (1998): Tanzania, Pangani, *Stuhlmann* I. 773 (B!, lecto.)

Branches smooth, grey, the youngest sometimes purple-brown. Petiole 5–15 mm. long; lamina fresh green to glaucous, thinly fleshy, ovate to broadly elliptic or obovate, 1.5–7 cm. long, 1–6 cm. wide, ± wavy at the margins, with 4–6 pairs of lateral nerves. Peduncle 0–3 mm. long; pedicels 1–4 mm. long; bract with a short ovate-triangular limb, 2–2.5 mm. long, ± ciliolate. Receptacle campanulate, 2–3 mm. long; calyx 0.8–2.5 mm. long. Corolla mostly pinkish red, but pale or greenish at the base, green to yellow around vents (through which red filaments are visible when buds are ripe), 6–8(–9) cm. long; tube 2–4 cm. long; lobes linear-spathulate, recurving about the middle. Stamen-filaments inserted 5–10 mm. long above the base of corolla-lobes, red; anthers 13–18 mm. long. Berry purple, obovoid, 8–12 mm. long; seed brilliant red.

SYN. *Loranthus panganensis* Engl. in E.J. 20: 92 (1894) & P.O.A. C: 165, t. 14H–L (1895); Sprague in F.T.A. 6(1): 316 (1910); F.D.O.-A. 2: 168 (1932); T.T.C.L.: 283 (1949); U.K.W.F.: 330, fig. on 333 (1974); Blundell, Wild Fl. Kenya: 66, t. 28/179 (1982)

subsp. **panganensis**; Polh. & Wiens, Mistletoes Afr.: 97 (1998)

Leaves usually produced at flowering time; petioles mostly 10–15 mm. long; lamina glaucous, broadly ovate-elliptic to obovate, 4.5–7 cm. long, 2–6 cm. wide, flat to somewhat undulate and sinuate. Calyx cupular, 2–2.5 mm. long, longer than receptacle. Corolla-lobes expanded at very base and again ± 5–6 mm. above where the filaments are inserted; mature corolla-buds thus with swellings above and below the vents, the lips of which are recurved and bright yellow.

KENYA. Northern Frontier Province: ± 10 km. N. of Marsabit, 4 Aug. 1979, *Bock* in E.A. 16429!; Machakos District: 29 km. SE. of Kibwezi, 24 Jan. 1972, *Wiens* 4547!; Kilifi District: Mariakani, 1929, *R.M. Graham* in F.D. 1818!
TANZANIA. Lushoto District: Mashewa [Mascheua], July [Aug. on sheet at K] 1893, *Holst* 3503!; Pangani, Dec. 1889, *Stuhlmann* I. 773!
DISTR. **K** 1, 4, 7; **T** 3; not known elsewhere
HAB. Deciduous and coastal bushland, generally on *Adansonia*, *Euphorbia* and *Sterculia*; 0–1250 m.

NOTE. Less conspicuous and often more difficult to collect than subsp. *commiphorae*.

subsp. **commiphorae** *Wiens & Polh.*, Mistletoes Afr.: 97, photo. 1, 26 (1998). Type: Kenya, Kwale District, 5 km. N. of Kinango, *Polhill* 4821 (K!, holo.)

Leaves developed after flowering; petiole 3–8 mm. long; lamina bright green, broadly elliptic, 1.5–4.5 cm. long, 1–3.5 cm. wide, crinkly. Calyx 1–1.5 mm. long, shorter than receptacle. Corolla-lobes narrow above expanded base, the filaments inserted 6–10 mm. from base; mature corolla-buds swollen only below the vents, the lips green or yellow-green like outside of the lobes. Fig. 3.

UGANDA. Karamoja District: Pirre, 10 Nov. 1939, *A.S. Thomas* 3256! & Katikekile, Oct. 1956, *J. Wilson* 286! & Lokitanyala, Sept. 1963, *Tweedie* 2725!
KENYA. Northern Frontier Province: Furrole [Furroli], 13 Sept. 1952, *Gillett* 13843!; Machakos District: Kiu, 15 Sept. 1961, *Polhill & Paulo* 460!; Kilifi District: 3 km. E. of Ganze, 15 Sept. 1985, *S.A. Robertson* 4037!
TANZANIA. Mbulu District: between Mto wa Mbu and Karatu, 5 km. from Manyara Hotel, 31 Mar. 1964, *Verdcourt* 4007!; Kahama District: 48 km. Nzega–Kahama, 26 July 1950, *Bullock* 3025!; Uzaramo District: creek near Dar es Salaam harbour, Feb. 1940, *Vaughan* 2945!
DISTR. **U** 1; **K** 1, 2, 4, 6, 7; **T** 1–4, 6, 7; Ethiopia, Somalia
HAB. Coastal and deciduous bushland, wooded grassland, generally on *Commiphora*, occasionally on *Lannea* or *Sterculia*; 0–1700 m.

FIG. 3. *EMELIANTHE PANGANENSIS* subsp. *commiphorae* — **1**, leafy shoot, × 1; **2**, inflorescence, × 1. Drawn by Eleanor Catherine. Reproduced from Flora of Ethiopia.

NOTE. This is the more commonly encountered subspecies. It has smaller bright green (rather than glaucous) markedly crinkled leaves mostly produced after the flowers. The filaments are inserted on the tapered part of the corolla-lobes about 7–10 mm. above their base. The mature corolla-buds have a single swelling at the top of the corolla-tube at the base of the vents. In subsp. *panganensis* the filaments are inserted on a second broadening of the corolla-lobes, about 5–6 mm. above the base. This is apparent in the mature buds as a second swelling, and when the vents open between them the recurved lips are bright yellow, contrasting more with the green to orange-yellow ground-colour of the lobe-bases and the bright red of the filaments inside. The calyx of subsp. *commiphorae* is also shorter. The geographical ranges overlap, but subsp. *panganensis* is rarely found on *Commiphora*, and generally on *Adansonia, Sterculia* or *Euphorbia*.

## 4. ONCOCALYX

Tiegh. in Bull. Soc. Bot. Fr. 42: 258 (1895); M.G. Gilbert, Polh. & Wiens in Nordic Journ. Bot. 5: 222 (1985); Polh. & Wiens, Mistletoes Afr.: 101 (1998)

*Loranthus* L. sect. *Dendrophthoë* (Mart.) Engl. group *Longicalyculati* Engl. in E.J. 20: 85 (1894), as '*Longecalyculati*'
*Odontella* Tiegh. in Bull. Soc. Bot. Fr. 42: 259 (1895); Balle in Bull. Séances Acad. Roy. Sci. Col., n.s., 2: 1072 (1956), *nom. illegit., non* C.A. Agardh (1832)

*Loranthus* L. subgen. *Tapinanthus* (Blume) Engl. sect. *Pentatapinanthus* Engl. group
*Coriaceifolii* Engl. in E. & P. Pf., Nachtr. 1: 133 (1897)
*L.* sect. *Longicalyculati* (Engl.) Sprague in F.T.A. 6(1): 266 (1910)
*L.* sect. *Coriaceifolii* (Engl.) Sprague in F.T.A.: 266 (1910)
*Danserella* Balle in Webbia 11: 583 (1955) & in Bull. Séances Acad. Roy. Sci. Col., n.s.,
2: 1069 (1956), *nom. provis.*
*Tieghemia* Balle in Bull. Séances Acad. Roy. Sci. Col., n.s., 2: 1062 (1956)

Small shrubs with a single haustorial connection, glabrous or with short spreading simple hairs. Leaves mostly alternate, with 1–2 pairs of strongly ascending nerves from near the base. Flowers clustered or 2–6 in very shortly pedunculate umbels; bract saucer-shaped to cupular, truncate or with a small limb. Calyx cupular to tubular. Corolla 5-merous, generally joined less than halfway, sometimes up to ²/₄ but then with long slits below filament-insertions, opening with a V-shaped split on one side, generally yellow at least in part, sometimes banded red and white or greenish, developing vents in mature buds; tube constricted above a basal swelling or narrow throughout; lobes linear, sometimes a little broadened above, sometimes dilated around vents at the base, erect, recurved or revolute. Filaments attached at the base of corolla-lobes or some distance above, inrolled at anthesis, sometimes with small appendages in front of the anther or a little below; anthers 4-thecous, truncate or with a small bilobed connective-appendage. Style filiform; stigma ovoid-globose to obovoid. Berry red, usually obovoid, often with a subpersistent calyx, occasionally verrucose.

13 species in the drier forests and bushland of eastern and southern Africa to Arabia.

*Oncocalyx* is generally easily recognised by the shortly tubular 5-merous flowers opening by a V-slit. In *O. sulfureus* (sect. *Oncocalyx*) the corolla is asymmetric only by a slight coherence of the erect corolla-lobes. In the other species (sect. *Longicalyculati* (Engl.) Polh. & Wiens) the filaments are consistently short and tightly coiled, and the fairly short narrow corolla-tube opens with a clear V-shaped split often almost down to the calyx. The corolla-lobes in this section may be erect or reflexed, the tube with or without a ring of small appendages inside, and the filaments with or without a small tooth in front of the anther. These features are species constant but occur in virtually all combinations.

The floral modifications in this genus set the pattern for the common pollination mechanism throughout most of the more advanced genera of Loranthaceae in Africa, although the result seems to have been obtained by convergence in several different groups. The principle features remain the development of vents in the mature buds with various signals that encourage birds to probe in search of nectar, thus splitting the corolla-tube with a V-slit, releasing the tension of the filaments and projecting the pollen on to the bird's head.

1. Corolla swollen at base; calyx shortly cupular (sect. *Oncocalyx*) · · · 1. *O. sulfureus*
   Corolla-tube narrow; calyx tubular (sect. *Longicalyculati*) · · · · · · · · · · · · · ·2
2. Corolla-lobes reflexed to revolute at anthesis: · · · · · · · · · · · · · · · · · · · · · ·3
   Corolla-lobes remaining erect · · · · · · · · · · · · · · · · · · · · · · · · · · · · · · · · ·4
3. Corolla-lobes expanded at base to form narrow roundly protruding pouches around vents of maturing buds; tube without appendages inside; petiole 2–6 mm. long; flowers arising separately (peduncle absent); plants subglabrous to densely hairy · · · · · · · · · · · · · · · · · · · · · · · · · · · 2. *O. fischeri*
   Corolla-lobes not broadened into pouches at the base; tube with a ring of small hairy appendages inside; petiole 0–1 mm. long; flowers in subsessile umbels; plants glabrous · · · 3. *O. ugogensis*
4. Leaves sessile, cordate, amplexicaul · · · · · · · · · · · · · · · · · 6. *O. cordifolius*
   Leaves petiolate, cuneate to rounded at base · · · · · · · · · · · · · · · · · · · ·5

5. Branchlets slightly compressed at first, soon subterete, glabrous to shortly hairy; leaves slightly but distinctly succulent, mostly broadest about the middle or below; calyx 4–6 mm. long, generally papillate to papillose-hairy ··   4. *O. kelleri**

Branchlets angular to winged, distinctly keeled below each node, glabrous; leaves thin, mostly broadest above middle; calyx 3–4 mm. long, smooth, glabrous ···············   5. *O. angularis*

1. **O. sulfureus** (*Engl.*) *Polh. & Wiens* in Lebrun & Stork, Énum. Pl. Fl. Afr. Trop. 2: 170 (1992); Blundell, Wild Fl. E. Afr.: 130 (1987), *nom. invalid.*; Polh. & Wiens in U.K.W.F., ed. 2: 155 (1994) & Mistletoes Afr.: 103, photo. 29, fig. 5A (1998). Type: Tanzania, Moshi District, Kilimanjaro, Useri, Tarakia [Karrakia] ravine, *Volkens* 2002 (B†, holo., BM!, iso., K!, fragment)

Plant glabrous; twigs compressed to slightly angular. Leaves alternate and clustered on short shoots; petiole 2–4 mm. long; lamina leathery, elliptic to elliptic-lanceolate or obovate, 1–4.5 cm. long, 0.6–2 cm. wide, bluntly pointed or rounded at the apex, basally attenuate into the petiole, with 1(–2) pairs of steeply ascending nerves from a little above the base. Flowers 2–4 in 1–several subsessile umbels in axils and at older nodes; pedicel 2–3 mm. long; bract obliquely ovate-cupular, keeled, 1.5–2 mm. long. Receptacle 1–1.5 mm. long; calyx shortly cupular, 0.5 mm. long, slightly toothed. Corolla 2.5–4 cm. long, yellow or greenish yellow, with the lobes red above filament-insertions, joined for one-third, the lobes remaining erect, with the tips slightly incurved and coherent; basal swelling 2.5–3 mm. long. Filaments inserted halfway up lobes, involute after anthesis, with a tiny rounded or bilobed flap jutting out just below anther; anther 2–3.5 mm. long, with a short connective-appendage. Berry bright red, obovoid, 9 mm. long, 7 mm. in diameter, slightly verrucose. Fig. 4/1–3.

KENYA. Nakuru District: Njoro R. valley, 21 Jan. 1985, *Kirkup* 15!; Nairobi District: Langata, near Hillcrest Secondary School, 1 Jan. 1980, *M.G. Gilbert* 5812!; Masai District: Nasampolai [Enesambulai] valley, 5 Mar. 1969, *Greenway & Kanuri* 13591!
TANZANIA. Moshi District: Kilimanjaro, Useri, 30 Mar. 1934, *Schlieben* 5009! & Useri, Tarakia [Karrakia] ravine, *Volkens* 2002! & above Leranjwa, 13 Nov. 1993, *Grimshaw* 93/1024!
DISTR. **K** 1, 3–6; **T** 2; not known elsewhere
HAB. Upland dry evergreen forest and associated bushland or wooded grassland; 1700–3000 m.

SYN. *Loranthus sulfureus* Engl. in P.O.A. C: 165 (1895); Sprague in F.T.A. 6(1): 338 (1910), as 'sulphureus'; F.D.O.-A. 2: 170 (1932); T.T.C.L.: 283 (1949), as 'sulphureus'; U.K.W.F.: 332 (1974), as 'sulphureus'
    *L. friesiorum* K. Krause in N.B.G.B. 8: 494, fig. 1A–D (1923). Lectotype, chosen by Polh. & Wiens (1998): Kenya, N. Nyeri District, Mt. Kenya, W. side, Forest Station, *R.E. & T.C.E. Fries* 463 (B!, lecto., BR!, K!, MO!, UPS, WAG!, isolecto.)
    *Tapinanthus friesiorum* (K. Krause) Danser in Verh. K. Akad. Wet., sect. 2, 29(6): 112 (1933)
    *T. sulfureus* (Engl.) Danser in Verh. K. Akad. Wet., sect. 2, 29(6): 120 (1933)
    *Tieghemia sulfurea* (Engl.) Balle in Bull. Séances Acad. Roy. Sci. Col., n.s., 2: 1066, fig. 4/11–31 (1956)

NOTE. *O. sulfureus* is the least specialised species of the genus. Its nearest relatives occur in South Africa.

2. **O. fischeri** (*Engl.*) *M.G. Gilbert* in Nordic Journ. Bot. 5: 222 (1985); Blundell, Wild Fl. E. Afr.: 129, t. 343 (1987); M.G. Gilbert in Fl. Ethiopia 3: 369, fig. 114.9 (1990); Polh. & Wiens in U.K.W.F., ed. 2: 155, t. 56 (1994) & Mistletoes Afr.: 106, photo. 32, fig. 5D (1998). Lectotype, chosen by Polh. & Wiens (1998): Kenya/Tanzania, Masailand, *Fischer* 130 (HBG, lecto., B!, isolecto.)

---

* *O. ugogensis* may be miskeyed here if the corolla-lobes have not yet reflexed, but the leaves are less succulent on very short petioles, the flowers are more definitely umbellate on short shoots, and the calyx is smooth, not papillate.

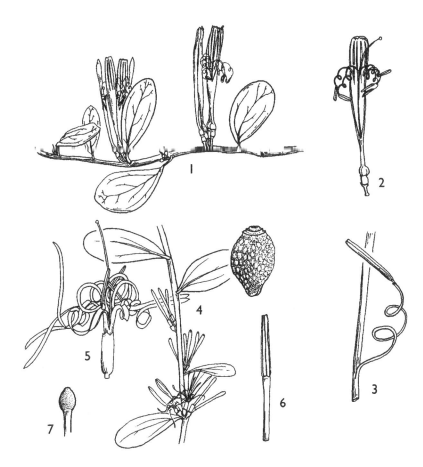

Fig. 4. *ONCOCALYX SULFUREUS* — 1, flowering branch, × ²/₃; **2**, flower, × 1; **3**, stamen, × 4. *ONCOCALYX FISCHERI* — **4**, flowering branch, × ²/₃; **5**, flower, × 2; **6**, stamen, × 4; **7**, style-tip, × 8; **8**, fruit, × 3. 1–3, from *Bally* 10699; 4–7, from *Tweedie* 2557; 8, from *Flora of Kenya* N94. Drawn by Christine Grey-Wilson.

Twigs almost glabrous to densely covered with short stiff spreading hairs. Leaves alternate; petiole 2–6 mm. long; lamina thinly coriaceous, grey-green to slightly glaucous, linear-lanceolate to elliptic, obovate or oblanceolate, 1–5.5 cm. long, 0.4–4 cm. wide, bluntly pointed to rounded at the apex, subglabrous to densely velvety-hairy on both surfaces, with 1–2 pairs of strongly ascending nerves from a little above the base or venation obscure. Flowers clustered in the axils or at nodes below; pedicel 0.5–2 mm. long; bract cupular, 2–3 mm. long, subglabrous to densely hairy with longer cilia. Receptacle 1–1.2 mm. long; calyx tubular, 4–6 mm. long, ciliolate. Corolla 2–2.7 cm. long, usually yellow fading orange, but sometimes red, opening by a unilateral split; tube narrow, 6–8 mm. long; lobes narrow, recurving and twisted to coiled after anthesis, expanded at the base to form pouches around vents of maturing buds. Filaments attached at the base of the lobes, involute and shedding anthers at anthesis. Berry red, obovoid-ellipsoid, 6–7 mm. long, 4–5 mm. in diameter, verrucose. Fig. 4/4–8.

UGANDA. Acholi District: Agoro, Mar. 1936, *Eggeling* 1745!; Karamoja District: Toror Hills, 8 Oct. 1952, *Verdcourt* 793! & base of Mt. Kadam [Debasien], May 1954, *Philip* 586!

Kenya. Northern Frontier Province: Malka Mari, 21 Jan. 1972, *Bally & Smith* 14920!; Masai District. Magadi road, E. of Kampi ya Bibi, 19 Feb. 1969, *Greenway & Napper* 13569!, Teita District: 8 km. W. of Bura Railway Station, 17 Jan. 1972, *Gillett* 19581!

Tanzania. Musoma District: Seronera R. at Seronera, 19 Apr. 1961, *Greenway & Myles Turner* 10073!; Arusha District: Ngare Nanyuki road, 15 Sept. 1969, *Richards* 24695!; Pare District: 3 km. N. of Same, 29 Dec. 1985, *Wiens* 6518!

Distr. **U** 1; **K** 1–7; **T** 1–5; Ethiopia and Somalia

Hab. Dry evergreen forest and higher rainfall bushland along Great Rift Valley and its flanks, extending down rivers almost to Kenya coast, commonly on *Acacia, Grewia, Cordia* and *Maytenus* but also on numerous other hosts; 30–2100 m.

Syn. *Loranthus fischeri* Engl. in E.J. 20: 85, t. 1A (1894) & P.O.A. C: 165 (1895); Sprague in F.T.A. 6(1): 331 (1910); Engl. in V.E. 3(1): 93, fig. 56 (1915); K. Krause in N.B.G.B. 8: 496 (1923); F.D.O.-A. 2: 169 (1932), pro majore parte; Engl. & Krause in E. & P. Pf., ed. 2, 16B: 162, fig. 80 (1935); Chiov., Racc. Bot. Miss. Consol. Kenya: 109 (1935); T.T.C.L.: 286 (1949); U.K.W.F.: 330, fig. on 327 (1974).

   *L. stuhlmannii* Engl. in E.J. 20: 85 (1894) & P.O.A. C: 165, t. 12A–C (1895); Sprague in F.T.A. 6(1): 331 (1910); F.D.O.-A. 2: 169 (1932); T.T.C.L.: 286 (1949); U.K.W.F.: 332 (1974). Type: Tanzania, Mwanza, *Stuhlmann* 4572 (B!, holo., K, P, fragments!)

   *L. stuhlmannii* Engl. var. *somalensis* Sprague in F.T.A. 6(1): 331 (1910) & in K.B. 1911: 145 (1911). Type: Ethiopia, Webi R. to Abdallah, *Keller* 218 (Z, holo., K!, iso., B!, fragment)

   *L. acacietorum* Bullock in K.B. 1931: 271 (1931); T.T.C.L.: 285 (1949); U.K.W.F.: 332 (1974). Type: Tanzania, Pare District, SW. Pare Mts., Mwembe, *Greenway* 2065 (K!, holo., EA!, iso.)

   *L. dichrostachydis* Chiov., Fl. Somala 2: 386, fig. 220 (1932). Type: Somalia, Licchitore, *Senni* 482 (FT, holo.)

   *Tapinanthus acacietorum* (Bullock) Danser in Verh. K. Akad. Wet., sect. 2, 29(6): 107 (1933)

   *T. fischeri* (Engl.) Danser in Verh. K. Akad. Wet., sect. 2, 29(6): 112 (1933)

   *T. stuhlmannii* (Engl.) Danser in Verh. K. Akad. Wet., sect. 2, 29(6): 120 (1933)

Note. The synonyms listed above indicate significant deviations from the typical facies of *O. fischeri*, but to give them formal rank is likely to be more confusing than helpful. The type of *Loranthus stuhlmannii* is from Mwanza on the S. side of Lake Victoria; like other specimens from around Lake Victoria the plants have sparse and evanescent hairs on the twigs and other parts and the leaves are quite exceptionally long and narrow. The rest of the material from Tanzania shows gradations to the typical form of *O. fischeri*, which has densely hairy branches, leaves and bracts and mostly elliptic to obovate leaves. There is a tendency for the southern outlying populations to have red or partly red flowers, as distinct from the yellow flowers of *O. fischeri* which turn to no more than orange as they fade. Specimens attributed to *Loranthus acacietorum* from the SE. limit of the range in the Chyulu and Pare Mts. also have subglabrous branchlets and bracts, but the leaves are broad and the old flowers are orange. Glabrous plants with reddish, sometimes rather large flowers (*L. dichrostachydis*) occur either side of the Somalia/Kenya border between the Juba and Tana rivers down to the coast.

3. **O. ugogensis** (*Engl.*) *Polh. & Wiens* in Nordic Journ. Bot. 5: 222 (1985); Blundell, Wild Fl. E. Afr.: 130, t. 344 (1987); Polh. & Wiens in U.K.W.F., ed. 2: 155 (1994) & Mistletoes Afr.: 107, photo. 33A–B (1998). Type: Tanzania, Dodoma District, Ugogo, Njassa [Mjesse], *Stuhlmann* 342 (B†, holo., K!, fragment)

Plant glabrous; twigs soon terete. Leaves alternate and clustered on short shoots; petiole 0–1 mm. long; lamina coriaceous, glaucous, elliptic-oblong to obovate, 1–5 cm. long, 0.3–3 cm. wide, bluntly pointed to rounded at the apex, with a pair of strongly ascending lateral nerves from a little above the base. Flowers 1–4 in subsessile umbels terminal on short shoots usually after the primary leaves have fallen; peduncle to 0.5 mm. long; pedicels 1 mm. long; bract cupular, with a short concave lobe, keeled, 2–3.5 mm. long, ± jagged or ciliolate. Receptacle 1–1.5 mm. long; calyx tubular, 5–6 mm. long, shortly lobed. Corolla 2.9–3.5 cm. long, yellow, greenish yellow or rarely red and yellow, opening by a unilateral split; tube 1–1.5 cm. long, with 5 small hairy flaps forming a ring inside; lobes slightly broadened upwards, reflexed from a little above the base after anthesis. Filaments attached at the base of the lobes, incurved and shedding anthers at anthesis, with a small tooth in front of the anther. Berry red, obovoid, 7 mm. long, 4 mm. in diameter.

UGANDA. Karamoja District: Moroto, 4 Jan. 1937, *A.S. Thomas* 2173!
KENYA. W. Suk District: N. of Marich Pass at foot of Kaimat Escarpment, 27 Oct. 1977, *Carter & Stannard* 79!; Baringo District: Lake Bogoria, 10 Oct. 1982, *S.A. Robertson* 3432!; Tana R. District: Galole, 17 Dec. 1964, *Gillett* 16375!
TANZANIA. Masai District: foot of Mt. Loolmalassin at Engaruka, 9 July 1956, *Bally* 10674!; Mpwapwa District: Kibakwe village, 10 Apr. 1988, *Bidgood, Mwasumbi & Vollesen* 991!; Kilosa District: Mbuyuni, 11 Apr. 1988, *Congdon* 202!
DISTR. U 1; K 1–4, 6, 7; T 2, 3, 5, 6; Somalia
HAB. Deciduous bushland, most commonly on *Acacia, Grewia* and Capparaceae, but also on *Cordia, Terminalia, Commiphora, Balanites*, etc.; 15–1450 m.

SYN. *Loranthus ugogensis* Engl. in E.J. 20: 86 (1894) & P.O.A. C: 165 (1895); Sprague in F.T.A. 6(1): 336 (1910); F.D.O.-A. 2: 170 (1932); T.T.C.L.: 286 (1949); U.K.W.F.: 332 (1974)
   *L. microphyllus* Engl. in E.J. 20: 86 (1894) & P.O.A. C: 165 (1895); Sprague in F.T.A. 6(1): 336 (1910); F.D.O.-A. 2: 170 (1932); T.T.C.L.: 286 (1949). Type: Kenya/Tanzania, probably Masailand, *Fischer* 311 (B!, holo., K!, fragment)
   *L. tanaensis* Engl. in E.J. 40: 528 (1908). Type: Kenya, Lamu District, Witu, *F. Thomas* 58 (B!, holo., BR!, COI!, K!, iso.)*
   *Tapinanthus microphyllus* (Engl.) Danser in Verh. K. Akad. Wet., sect. 2, 29(6): 116 (1933)
   *T. ugogensis* (Engl.) Danser in Verh. K. Akad. Wet., sect. 2, 29(6): 121 (1933)


4. **O. kelleri** (*Engl.*) *M.G. Gilbert* in Nordic Journ. Bot. 5: 222 (1985) & in Fl. Ethiopia 3: 370 (1990); Polh. & Wiens, Mistletoes Afr.: 109, photo. 35 (1998). Type: Ethiopia, Webi R. to Abdallah, *Keller* 219 (Z, holo., B!, K!, fragments)

Twigs slightly compressed, soon terete, glabrous to very shortly spreading hairy. Leaves alternate to subopposite; petiole 2–7 mm. long; lamina fleshy, slightly glaucous, ovate to elliptic-oblong or oblanceolate, 2–7(–9) cm. long, 1–4.5 cm. wide, bluntly pointed to rounded at the apex, cuneate to rounded at the base, glabrous to shortly spreading hairy, with 1–2 pairs of strongly ascending lateral nerves from near the base. Flowers clustered in the axils or more commonly at older nodes; pedicels 1–3 mm. long; bract cupular, 2–4 mm. long, subtruncate or with a short lobe, slightly keeled or gibbous, glabrous to papillose-hairy. Calyx tubular, 4–6 mm. long, slightly lobed, papillose. Corolla 3–3.5 cm. long, orange-yellow, sometimes green to red on tube and red on lobes, glabrous to papillose-hairy, opening by a unilateral split down to calyx, tube 1.1–1.4 cm. long, with a ring of 5 hairy appendages just below the V-split; lobes remaining erect, linear. Stamen-filaments attached at the base of the lobes, coiling inwards and shedding anthers at anthesis, with a small tooth in front of the anther. Berry red, ellipsoid to obovoid, 8–10 mm. long, 5–7 mm. in diameter, excluding the subpersistent calyx, glabrous to papillose-hairy; seed orange-yellow.

KENYA. Northern Frontier Province: R. Kargi at Kargi, 12 June 1960, *Oteke* 167!; Meru District: Isiolo, 9 Mar. 1952, *Gillett* 12511!; Tana R. District: Kurawa, 20 Sept. 1961, *Polhill & Paulo* 500!
TANZANIA. Lushoto District: ± 13 km. N. of Mkomazi on road to Gonja and Kisiwani, 30 Dec. 1985, *D. & C. Wiens* 6521! & 8 km. SE. of Mkomazi, 30 Apr. 1953, *Drummond & Hemsley* 2297!; Kilosa District: Malolo to Mbuyuni, 12 Apr. 1988, *Lovett & Congdon* 3245!
DISTR. K 1, 4, 7; T 3, 6; Ethiopia and Somalia
HAB. Deciduous bushland, commonly riverine, usually on Capparaceae or Salvadoraceae; 15–1100 m.

SYN. *Loranthus kelleri* Engl. in E.J. 40: 537 (1908); Sprague in F.T.A. 6(1): 337 (1910)
   *L. kihuirensis* Engl. in E.J. 40: 528 (1908); Sprague in F.T.A. 6(1): 334 (1910); F.D.O.-A. 2: 170 (1932); T.T.C.L.: 286 (1949). Type: Tanzania, Pare District, Kihuiro–Gonja, *Engler* 1541 (B, holo., K!, fragment)
   [?*L. fischeri* Engl. var. *glabratus* sensu F.D.O.-A. 2: 170 (1932), quoad *Peter* 41320, *non* Engl.]
   *Tapinanthus kelleri* (Engl.) Danser in Verh. K. Akad. Wet., sect. 2, 29(6): 114 (1933)
   *T. kihuirensis* (Engl.) Danser in Verh. K. Akad. Wet., sect. 2, 29(6): 114 (1933)
   [?*L. glabratus* sensu T.T.C.L.: 286 (1949), quoad *Peter* 41320, *non* (Engl.) Sprague]

* The specimen at BM is *O. fischeri*.

5. **O. angularis** *M.G. Gilbert* in Nordic Journ. Bot. 5: 222 (1985) & in Fl. Ethiopia 3: 370 (1990); Polh. & Wiens, Mistletoes Afr.: 110, photo. 36 (1998). Type: Kenya, Northern Frontier Province, Kowop, *M.G. Gilbert, Gachathi & Gatheri* 5172 (K!, holo., EA!, ETH, iso.)

Plant glabrous; twigs angular, keeled below each node. Leaves alternate; petiole 2–3 mm. long; lamina fleshy, glaucous or grey, ovate-elliptic to elliptic-oblong or slightly obovate, 1–3.5 cm. long, 0.4–2.5 cm. wide, bluntly pointed to rounded at the apex, cuneate at the base, with a pair of strongly ascending lateral nerves from near the base or venation obscure. Flowers clustered in axils or at nodes below; pedicels 1–3 mm. long; bract cupular with a distinct lobe, keeled below it, 2–3 mm. long. Receptacle 1 mm. long; calyx tubular, 3–4 mm. long, slightly lobed, sometimes slightly pinkish. Corolla 2.8–3.3 cm. long, yellow in a zone at the base of the lobes, variously green to yellow-green above and below, opening by a unilateral split nearly to the calyx; tube 1–1.2 cm. long, with a ring of 5 small hairy appendages just below the V-slit inside; lobes linear, remaining erect after anthesis. Filaments attached at the base of the lobes, coiling inwards and shedding anthers at anthesis, with a well-developed tooth in front of the anther. Berry not seen.

KENYA. Northern Frontier Province: Loriu Plateau, 3 June 1970, *Mathew* 6608! & Laisamis, 14 June 1960, *Oteke* 177! & Gara Faiyu [Faio], 14 Mar. 1952, *Gillett* 12535!
DISTR. **K** 1; S. Ethiopia and S. Somalia
HAB. Deciduous bushland, on a variety of hosts including *Euphorbia*; 50–1500 m.

6. **O. cordifolius** *Wiens & Polh.*, Mistletoes Afr.: 110, photo. 37 (1998). Type: Kenya, Lamu District, Manda I., Takwa, *Greenway* 9439 (K!, holo., EA!, iso.)

Plant glabrous; twigs angular to slightly winged. Leaves alternate, sessile, rather fleshy, glaucous, cordate, 3–5 cm. long, 2–4 cm. wide, amplexicaul, with 1–2 pairs of strongly ascending lateral nerves from near the base. Flowers clustered in the axils; pedicels 2–3 mm. long; bract shortly cupular, lobed and keeled below, 2 mm. long. Receptacle 1 mm. long; calyx tubular, 3–4 mm. long, subtruncate to slightly lobed. Corolla 3–3.3 cm. long, yellow, fading yellow-orange, distal part of the lobes sometimes orange, opening by a unilateral split; tube 1.2–1.3 cm. long, with a ring of 5 small downwardly directed shortly hairy flaps just below the V-split inside; lobes linear, remaining erect after anthesis. Stamen-filaments attached to the base of the lobes, coiling inwards and shedding anthers at anthesis, with a distinct tooth in front of the anther. Berry red, obovoid, 8 mm. long, 6 mm. in diameter, excluding the persistent calyx.

KENYA. Lamu District: Mvundeni to Stesheni, 26 Nov. 1988, *S.A. Robertson & Luke* 5559! & Ishakani ruins just N. of Shakani on coast between Kiunga and Dar es Salaam, 5 Apr. 1980, *M.G. Gilbert & Kuchar* 5871! & Manda I., 2 Jan. 1976, *Bock in E.A.* 16032!
DISTR. **K** 7; coast of S. Somalia
HAB. Coastal bushland, on various hosts; near sea-level

## 5. SPRAGUEANELLA

Balle in Bull. Séances Acad. Roy. Sci. Col. 25: 1632 (1954) & n.s., 2: 1078, fig. 3 (1956); Polh. & Wiens, Mistletoes Afr.: 112 (1998)

*Loranthus* L. sect. *Rhamnifolii* Sprague in F.T.A. 6(1): 267 (1910)

Small shrubs with a single haustorial connection, glabrous. Leaves subopposite, shortly petiolate; nerves few, spreading to strongly ascending, sometimes obscure. Flowers 2–6 in shortly pedunculate axillary umbels; bract shortly cupular, with a small limb, gibbous or spurred dorsally. Calyx cupular to shortly tubular, split by expanding corolla. Corolla straight or bent in bud, 5-merous with a short tube, yellow to red, banded green or white, developing vents in mature buds; lobes linear-oblanceolate, often somewhat coherent, but not around a V-split in the tube, reflexed about the middle. Filaments attached at the base of corolla-lobes, erect and filiform below, upper part sometimes slightly thickened and inrolled at anthesis, with a small tooth in front of the anther; anthers linear, 4-thecous, truncate. Style filiform; stigma capitate. Berry obovoid, not seen mature.

Two species in East and South-Central Africa along coast and extending into mountains in drier types of forest.

Closely related to *Oncocalyx*, but with a very short corolla-tube without a V-split on one side; the common species, *S. rhamnifolia*, has characteristic flower-buds bent near the base.

Calyx saucer-shaped, 0.5–1 mm. long; corolla-tube flask-shaped;
    buds bent ± at right-angles above basal swelling; filaments
    filiform below, thickened and involute above · · · · · · · · ·   1. *S. rhamnifolia*
Calyx shortly tubular, 1.5–2.5 mm. long; corolla-tube narrow;
    buds straight or slightly bent; filaments uniform · · · · · · ·   2. *S. curta*

1. **S. rhamnifolia** (*Engl.*) *Balle* in Bull. Séances Acad. Roy. Sci. Col. 25: 1634 (1954) & n.s., 2: 1078, fig. 3 (1956); Vollesen in Opera Bot. 59: 64 (1980); Polh. & Wiens, Mistletoes Afr.: 112, photo. 38, fig. 6 (1998). Lectotype, chosen by Balle (1954): Tanzania, Tanga District, Amboni, *Holst* 2796 (B!, lecto., K!, isolecto.)

Stems well branched to 1 m. long; twigs subangular, soon densely lenticellate. Petiole 1–3 mm. long; lamina slightly yellow-green, thinly coriaceous, elliptic to oblong-elliptic or ovate, 4–10.5 cm. long, 1.5–6 cm. wide, bluntly pointed at the apex, basally attenuate, with 1–3 pairs of strongly ascending nerves from above the base. Flowers 2–6 in several shortly pedunculate umbels in the axils and at old nodes; peduncle 1–3 mm. long; pedicels 1–2 mm. long; bract shortly cupular, with a spurred ovate-triangular limb, 1–1.5 mm. long. Receptacle 1.5–2 mm. long; calyx 0.5–1 mm. long, lobed. Corolla bent in bud, 1.5–2 cm. long, red at the base, then greenish yellow or white shading to pink or red above; tube flask-shaped, 3–4 mm. long; lobes linear, broadening to linear-lanceolate at tips, reflexing below middle at anthesis and sometimes slightly coherent above. Filaments filiform below, thickened and at anthesis involute above. Fig. 5 (p. 22).

KENYA. Kitui District: Nuu Hill, 4 Mar. 1960, *Bally* 12093!; Kwale District: Mrima Hill, 4 Sept. 1957, *Verdcourt* 1864!; Kilifi District: 20 km. W. of Malindi on road to Jilore, 16 Jan. 1972, *Wiens* 4525!
TANZANIA. Tanga District: Machui on Tanga–Pangani road, 2 June 1955, *Faulkner* 1641!; Kilosa District: Mkwakwajuni, 27 June 1973, *Greenway & Kanuri* 15257!; Morogoro District: Dindili Hill, 30 June 1983, *Polhill, Lovett & Hall* 4951!
DISTR. **K** 4, 7; **T** 3, 6, 8; Malawi, Mozambique and E. Zimbabwe
HAB. Forest along coast, inland along rivers, in drier mist forest on hilltops and in higher rainfall wooded grassland; 0–1500 m.

SYN. *Loranthus rhamnifolius* Engl. in E.J. 20: 87 (1894) & P.O.A. C: 165, t. 12D–H (1895); Sprague in F.T.A. 6(1): 339 (1910); F.D.O.-A. 2: 171 (1932); T.T.C.L.: 283 (1949)
    *Oncocalyx rhamnifolius* (Engl.) Tiegh. in Bull. Soc. Bot. Fr. 42: 258, 740 (1895), *nom. invalid., comb. non rite publ.*
    ?*L. rhamnifolius* Engl. forma *grandifolius* Engl. in F.D.O.-A. 2: 155, 171 (1932); T.T.C.L.: 283 (1949), *nomen*. Based on Tanzania, Uzaramo District, Dar es Salaam, *Holtz* 546 (B)
    *Tapinanthus rhamnifolius* (Engl.) Danser in Verh. K. Akad. Wet., sect. 2, 29(6): 118 (1933)

2. **S. curta** *Wiens & Polh.*, Mistletoes Afr.: 112 (1998). Type: Tanzania, Lushoto District, W. Usambara Mts., Shagai Forest, *Drummond & Hemsley* 2577 (K!, holo., B!, BR!, SRGH!, iso.)

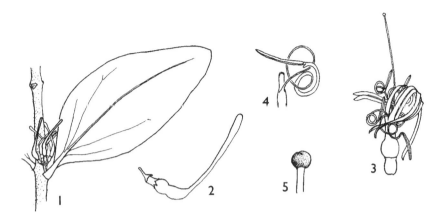

FIG. 5. *SPRAGUEANELLA RHAMNIFOLIA* — **1**, branch with young inflorescence, × ²/₃; **2**, flower bud, × 2; **3**, flower, × 2; **4**, stamen, × 4; **5**, style-tip, × 12. From *Gillespie* 377. Drawn by Christine Grey-Wilson.

Stems much branched; twigs subangular. Petiole 1–3 mm. long; lamina coriaceous, elliptic-oblanceolate to elliptic-obovate, 1.5–4.5 cm. long, 0.8–3 cm. wide, rounded to apex, attenuate at the base, lateral nerves spreading, obscure. Flowers 2–4 in axillary umbels; peduncle and pedicels ± 1 mm. long; bract shortly cupular, 1.5–2 mm. long, jagged, slightly gibbous. Receptacle 1.5 mm. long; calyx shortly tubular, 1.5–2.5 mm. long, slightly lobed, ciliolate, sometimes split by expanding corolla. Corolla 1.2–1.3 cm. long, dull red, paler towards the base, straight or only slightly bent in bud; tube narrow, 3 mm. long; lobes linear, broadening to linear-lanceolate at tips, reflexing at anthesis and sometimes slightly coherent. Filaments linear, involute, 3 times as long as anther.

TANZANIA. Lushoto District: Shagai Forest near Sunga, 17 May 1953, *Drummond & Hemsley* 2577! & between Ndabwa village and Lukosi, 31 Dec. 1985, *Wiens* 6529! & Shume Forest ± 3 km. SE. of Magamba road junction, 31 Dec. 1985, *Wiens* 6533!
DISTR. **T** 3; known only from the W. Usambara Mts.
HAB. Upland evergreen forest, on *Drypetes, Podocarpus* and introduced *Acacia*; 1800–1950 m.

## 6. OLIVERELLA

Tiegh. in Bull. Soc. Bot. Fr. 42: 258 (1895); Polh. & Wiens, Mistletoes Afr.: 114 (1998)

*Loranthus* L. sect. *Dendrophthoë* (Mart.) Engl. group *Involutiflori* Engl. in E.J. 20: 95 (1894)
L. sect. *Involutiflori* (Engl.) Sprague in F.T.A. 6(1): 261 (1910)

Small shrubs, with a single haustorial attachment; hairs short, simple, spreading. Leaves opposite to alternate, with lower nerves generally more ascending. Umbels solitary in axils, pedunculate, many-flowered; bract saucer-shaped, with subulate to foliaceous limb. Calyx rim-like. Corolla 5-merous, hairy, opening down one side; buds slightly curved, with ellipsoid apical swelling; tube green, short, with internal projections from infold at sinuses between lobes or flaps halfway down; lobes red, sometimes grey outside on upper part, separate below, framing vents of mature bud, coherent above, free at tips, inrolled at anthesis. Filaments attached to corolla-lobes, filiform below, thickened above, inrolled; anthers 4-thecous, with small connective-appendage. Style filiform; stigma capitate. Berry red, obovoid, with persistent calyx.

Three species in E. and S. central Africa in coastal and deciduous bushland and mixed woodland.

*Oliverella* is most similar to *Oncocalyx*, but easily recognised by the corolla-lobes, which are joined above and inrolled when the flower opens.

Twigs, leaves, umbels and corolla conspicuously covered in
    short spreading hairs · · · · · · · · · · · · · · · · · · · · · · · ·    1. *O. hildebrandtii*
Twigs, etc. puberulous but soon glabrescent · · · · · · · · · · · ·    2. *O. bussei*

1. **O. hildebrandtii** (*Engl.*) *Tiegh.* in Bull. Soc. Bot. Fr. 42: 259 (1895); M.G. Gilbert in Fl. Ethiopia 3: 371, fig. 114.11 (1990); Polh. & Wiens in U.K.W.F., ed. 2: 155 (1994) & Mistletoes Afr.: 115, photo. 40A–C, fig. 8 (1998). Lectotype, chosen by Polh. & Wiens (1998): Kenya, Teita District, Ndi, *Hildebrandt* 2852 (B!, lecto., BM!, isolecto.)

Branchlets velvety pubescent. Leaves opposite to alternate; petiole 1–12 mm. long; lamina thinly coriaceous, green to grey-green, lanceolate to elliptic or ovate, 3–8 cm. long, 1–4.5 cm. wide, narrowed or slightly acuminate to acute or obtuse at the apex, cuneate at the base, pubescent on both surfaces, somewhat glabrescent, with 4–10 pairs of lateral nerves, the lower ones sometimes stronger and more ascending. Umbels 8–20-flowered, hairy; peduncle 1–3 mm. long; pedicels 1–5 mm. long; bract with a subulate limb (2–)3–8 mm. long, sometimes long-spurred. Receptacle obconic to urceolate, 1.5–2 mm. long; calyx 0.2 mm. long. Corolla 1.4–1.7 cm. long (before lobes inrolled), hairy; tube yellow, greenish or white, 4–5 mm. long, with 5 V-shaped flaps halfway down; lobes free in lower third, red, the upper joined parts grey outside and red inside, inrolled above. Filaments attached to coherent part of the lobes, short; anthers 1.5–2 mm. long. Style green; stigma capitate, 0.5–0.6 mm. across. Berry red, sometimes speckled white, obovoid, 7–8 mm. long, 5–6 mm. in diameter, glabrescent; seed orange. Fig. 6 (p. 24).

UGANDA. Karamoja District: Turkana Escarpment, E. of Loyoro, July 1954, *Philip* 642! & Kidepo National Park, 7 July 1967, *Harrington* 313!
KENYA. Northern Frontier Province: Moyale, 23 Apr. 1952, *Gillett* 12901!; Embu District: 5 km. W. of Ishiara, 25 May 1975, *S.A. Robertson* 2215!; Kwale District: S. end of Mwachi Forest near bridge across Mwachi R., 17 Feb. 1977, *Faden, Gillett & Gachathi* 77/471!
TANZANIA. Handeni District: Mgambo, June 1957, *Mgaza* 147!; Kilosa District: Vuma Hill, 9 July 1973, *Greenway & Kanuri* 15379!; Uzaramo District: Kunduchi, 1 Aug. 1970, *B.J. Harris & Mwasumbi* 4917!
DISTR. U 1; K 1, 2, 4, 6, 7; T 3, 5, 6, ?8; S. Ethiopia and possibly S. Sudan, N. Mozambique
HAB. Deciduous and coastal bushland, commonly on *Grewia*, but also various other hosts, mostly in Combretaceae, Euphorbiaceae and Leguminosae; 0–1800 m.

SYN. *Loranthus hildebrandtii* Engl. in E.J. 20: 96 (1894) & P.O.A. C: 165, t. 15J–L (1895); Sprague in F.T.A. 6(1): 299 (1910); Engl. in V.E. 3(1): 94, fig. 57 (1915); K. Krause in N.B.G.B. 8: 493 (1923); F. D.O.-A. 2: 163 (1932); Engl. & K. Krause in E. & P. Pf., ed. 2, 16B: 155, fig. 74 (1935); T.T.C.L.: 286 (1949); U.K.W.F.: 330 (1974)
    *L. campestris* Engl. in E.J. 20: 95 (1894) & P.O.A. C: 165, t. 15E–H (1895); Sprague in F.T.A. 6(1): 300 (1910); F.D.O.-A. 2: 163 (1932); T.T.C.L.: 286 (1949). Type: Tanzania, Lushoto District, Mashewa, Kumba valley, *Holst* 3504 (B†, holo., K, fragment)
    *Oliverella sacleuxii* Tiegh. in Bull. Soc. Bot. Fr. 42: 259 (1895). Type: Tanzania, "Zanguebar", 1889, *Sacleux* 878 (P, holo.)
    *O. campestris* (Engl.) Tiegh. in Bull. Soc. Bot. Fr. 42: 259 (1895)
    *Loranthus orientalis* Engl. in E. & P. Pf., Nachtr. 1: 131 (1897), *non L. sacleuxii* (Tiegh.) Engl. (1897). Based on *O. sacleuxii* Tiegh.
    *Tapinanthus campestris* (Engl.) Danser in Verh. K. Akad. Wet., sect. 2, 29(6): 109 (1933)
    *T. hildebrandtii* (Engl.) Danser in Verh. K. Akad. Wet., sect. 2, 29(6): 113 (1933)
    *T. sacleuxii* (Tiegh.) Danser in Verh. K. Akad. Wet., sect. 2, 29(6): 119 (1933)

NOTE. See note under *O. bussei*.

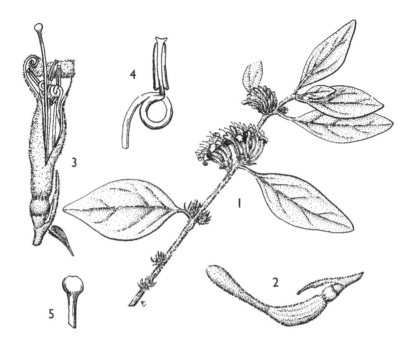

FIG. 6. *OLIVERELLA HILDEBRANDTII* — **1**, flowering branch, × ²/₃; **2**, flower bud, × 3; **3**, flower, × 3; **4**, stamen, × 5; **5**, style-tip, × 8. 1, from *M.G. Gilbert* 5009; 2–5, from *Napier-Bax* TNP/GS/20. Drawn by Eleanor Catherine. Reproduced from Flora of Ethiopia.

2. **O. bussei** (*Sprague*) *Polh. & Wiens* in Lebrun & Stork, Énum. Pl. Fl. Afr. Trop.: 169 (1992) & Mistletoes Afr.: 117 (1998). Lectotype, chosen by Polh. & Wiens (1998): Tanzania, Lindi District, Nangaru [Namgara], *Busse* 2949 (B!, lecto., BR!, EA!, iso., K!, fragment)

Branchlets puberulous, soon glabrescent. Leaves opposite; petiole 5–10 mm. long; lamina thinly coriaceous, lanceolate or elliptic, 3–6.5 cm. long, 1–2.5 cm. wide, slightly acuminate to obtuse at the apex, cuneate at the base, soon glabrescent, with oblique lateral nerves. Umbels 12–16-flowered, sparingly hairy; peduncle ± 2 mm. long; pedicels 1–1.5 mm. long; bract linear-lanceolate, long-spurred, 2–4 mm. long, ciliolate. Receptacle 0.8–1 mm. long; calyx 0.2–0.3 mm. long. Corolla pink, 1.3–1.5 cm. long (before lobes inrolled); tube 3 mm. long, with 5 V-shaped flaps halfway down; lobes red, free for ± 4 mm., ciliate, the upper parts joined. Filaments attached to coherent part of the lobes, short; anthers 1.5–2 mm. long. Stigma 0.6 mm. across. Berry red, apparently similar to *O. hildebrandtii*.

TANZANIA. Lindi District: Tendaguru, 26 May 1929, *Migeod* 524! & Lindi Creek, 15 July 1995, *Clarke* 88!; Mikindani, 1 June 1935, *Schlieben* 6546!
DISTR. **T** 8; not known elsewhere
HAB. Coastal bushland and wooded grassland; 0–200 m.

SYN. *Loranthus bussei* Sprague in F.T.A. 6(1): 299 (1910) & in K.B. 1911: 136 (1911); F.D.O.-A. 2: 163 (1932); T.T.C.L.: 285 (1949)
  *Tapinanthus bussei* (Sprague) Danser in Verh. K. Akad. Wet., sect. 2, 29(6): 109 (1933)

NOTE. As more material becomes available it seems increasingly likely that *O. bussei* is no more than a local variant of *O. hildebrandtii*. It is distinctly less hairy, but there do not seem to be any other significant differences apparent from dried specimens. The specimens from northern Mozambique have been provisionally attributed to *O. hildebrandtii*, but should not be excluded from a re-evaluation of *O. bussei*. Peter (1932) cites *Holtz* 639 from Lindi, Ras Rangone, as *O. hildebrandtii*, but this specimen may well be from Ras Rongoni point in Uzaramo District or misidentified. Peter also mentions the apparently unpublished name *Loranthus hildebrandtii* var. *glabrescens* Engl. under *O. bussei*, also based on *Busse* 2949 (B), and this name might be taken up.

## 7. ENGLERINA

Tiegh. in Bull. Soc. Bot. Fr. 42: 257 (1895); Polh. & Wiens, Mistletoes Afr.: 119 (1998)

*Loranthus* L. sect. *Ischnanthus* Engl. in E.J. 20: 125 (1894); Sprague in F.T.A. 6(1): 272 (1910)
*Ischnanthus* (Engl.) Tiegh. in Bull. Soc. Bot. Fr. 42: 260 (1895)
L. subgen. *Dendrophthoë* (Mart.) Engl. subsect. *Diplobracteati* Engl. in E.J. 40: 522 (1908)
L. sect. *Diplobracteati* (Engl.) Sprague in F.T.A. 6(1): 273 (1910)

Shrubs 0.5–2 m., with a single haustorial attachment, glabrous or shortly hairy; hairs simple or slightly branched, spreading, with an understorey of minute irregular trichomes, often giving a scurfy appearance. Leaves opposite or subopposite, the lowermost on flowering branchlets sometimes intergrading with bud-scales, sessile to generally petiolate; lamina sometimes coriaceous, but characteristically thin, sometimes flushed coppery brown, penninerved, but generally 1–several lower nerves more ascending. Flowers in pedunculate umbels, 2–20, often standing up like candles from horizontal branches; bract unilaterally developed from a generally shallow cupular base, not or scarcely exceeding the calyx, often slightly gibbous. Calyx annular or cup-shaped, sometimes lacerate or irregularly 4-lobed. Corolla 4-lobed, joined for one-fifth to two-thirds, red, yellow, orange or pink and white, often darker on bud-swellings; buds generally 4-angular, with vents marked by bosses, often with a slight to distinctly flask-shaped basal swelling, generally at least slightly swollen over the anthers, apiculate, obtuse, truncate or 4-bossed at the tip; lobes erect, reflexed or revolute, linear to distinctly spathulate above, sometimes hardened and shiny on the inner face of the dilated part, sometimes widened at the base around vents; tube split unilaterally, the V-slit extending halfway to almost the base, often papillate inside, at least along the sutures. Stamens attached at or near the base of corolla-lobes, the adnate part sometimes swollen or continued as a marked ridge down the corolla-tube, the free part tapered upwards, involute or inflexed at anthesis, produced into a tooth in front of the anther; anthers 4-thecous, with the connective slightly produced, minutely bifid. Style filiform, 4-angular; stigma obovoid to globular. Berry urceolate to obovoid, with persistent calyx, generally red; seed, where known, orange or yellow.

25 species in tropical Africa.

Technically *Englerina* differs from *Agelanthus* only by the 4-merous flowers, but the pattern of divergence is different and the commonly encountered species in eastern Africa are generally characterised also by the short corolla-tube and often thin leaves. The short corolla-tube in the relatively advanced species provides a similarity to *Oncocalyx* and is probably at least in part a convergence related to the influence of short-billed pollinators. The corolla develops vents like *Oncocalyx* and *Agelanthus*, but the opening may be facilitated in some cases, as observed in *E. inaequilatera*, by squeezing the bud open.

1. Corolla-lobes erect or slightly spreading · · · · · · · · · · · · · · · · · · · · · · · · · · 2
   Corolla-lobes reflexed to revolute · · · · · · · · · · · · · · · · · · · · · · · · · · · · · · 9
2. Corolla-tube 2–3.5 cm. long, longer than lobes · · · · · · · · · · · · · · · · · · · · · 3
   Corolla-tube up to 1.6 cm. long, shorter than lobes · · · · · · · · · · · · · · · · · · 4

3. Pedicels forming cupular sockets; bract cupular; calyx
   tubular, 2–3 mm. long · · · · · · · · · · · · · · · · · · · · · ·      1. *E. kwaiensis*
   Pedicels slender, 1.5–6 mm. long; bract saucer-shaped,
   with a small limb; calyx cupular, 0.8–1 mm. long · · · ·      2. *E. longiflora*
4. Leaves sessile, cordate · · · · · · · · · · · · · · · · · · · · · · · ·      7. *E. cordata*
   Leaves petiolate · · · · · · · · · · · · · · · · · · · · · · · · · · · · · · · · · · · · · 5
5. Corolla 9–12 mm. long, yellow tipped orange; tube
   smooth inside · · · · · · · · · · · · · · · · · · · · · · · · · · ·      8. *E. triplinervia*
   Corolla (11–)15–40 mm. long, if less than 15 mm. then
   tube grey outside and blue-black inside; tube
   conspicuously papillose to verruculose inside · · · · · · · · · · · · · · · · · 6
6. Inflorescence-axes and bracts persistently tomentellous
   with brown and white hairs; corolla-tube yellow inside,
   stamens green, darker and brown-spotted at base · · ·      6. *E. muerensis*
   Inflorescence-axes glabrous to puberulous; corolla-tube
   yellow to orange to red or blue-black inside, filaments
   at least dark at base · · · · · · · · · · · · · · · · · · · · · · · · · · · · · · · · · · 7
7. Calyx (1–)1.5–4 mm. long; receptacle in flower 1.5–2.5
   mm. long; basal swelling of corolla-buds (unless insect-
   galled) scarcely apparent; young twigs and inflorescence-
   axes usually scurfy-puberulous; corolla orange or
   yellow-orange, yellowish inside tube · · · · · · · · · · · · ·      3. *E. schubotziana*
   Calyx 0.5–1 mm. long; receptacle in flower 0.5–1.5 mm.
   long; basal swelling of corolla-buds flask-shaped
   (dilated a little above base and gradually narrowed to
   vents); twigs and inflorescence-axes glabrous or
   sometimes (in *E. inaequilatera*) puberulous · · · · · · · · · · · · · 8
8. Corolla (1.1–)1.5–2.5 cm. long, blue-grey with red lobes;
   tube blue-black inside, papillate principally along the
   sutures · · · · · · · · · · · · · · · · · · · · · · · · · · · · · · · · · · ·      4. *E. woodfordioides*
   Corolla 2.5–4 cm. long, yellow, orange or red, tube
   orange or dark red inside, with generally scattered
   papillae inside · · · · · · · · · · · · · · · · · · · · · · · · · · · · ·      5. *E. inaequilatera*
9. Corolla 3.2–4.5 cm. long, yellow or orange yellow, ciliate
   along tube-sutures and lobe-edges or hairy overall · · ·      14. *E. heckmanniana*
   Corolla 1.3–2.6 cm. long, glabrous · · · · · · · · · · · · · · · · · · · · · · · · · 10
10. Corolla-lobes white, greyish or greenish with pink to red
    tips; filaments 11–16 mm. long, loosely coiled above
    (measure mature corolla-bud from vents to base of
    apical swelling) · · · · · · · · · · · · · · · · · · · · · · · · · · · · · · · · · · · · · · 11
    Corolla-lobes red, yellow or orange, rarely white below
    (*E. macilenta*); filaments 5–9 mm. long, tightly coiled · · · · · · · · · · · · · 12
11. Corolla-tube generally red, smooth inside; filaments
    slightly dilated at point of attachment; flowers 8–16 per
    umbel, on pedicels 4–8 mm. long; calyx 0.5–1.5 mm.
    long · · · · · · · · · · · · · · · · · · · · · · · · · · · · · · · · · · · ·      9. *E. holstii*
    Corolla-tube pale, hispidulous papillate along sutures in
    upper half inside; filaments not dilated at base;
    flowers 4–6, on pedicels 3–5 mm. long; calyx 0.3–0.5
    mm. long · · · · · · · · · · · · · · · · · · · · · · · · · · · · · ·      10. *E. kagehensis*
12. Filaments not dilated at base, tapered into corolla-tube;
    anthers 2.5–3 mm. long; berry pear-shaped to almost
    obconic · · · · · · · · · · · · · · · · · · · · · · · · · · · · · · ·      13. *E. drummondii*
    Filaments slightly dilated at base, arising from base of
    corolla-lobes; anthers 4–6.5 mm. long; berry obovoid
    (unknown in *E. macilenta*) · · · · · · · · · · · · · · · · · · · · · · · · · · · · · 13

13. Anthers 5.5–6.5 mm. long; tooth 1 mm. long below
    anther; corolla-tube smooth outside; leaves obtuse · · ·    11. *E. ramulosa*
    Anthers 4–5 mm. long; tooth 0.5 mm. long, inserted 0.5
    mm. below anther; corolla-tube corrugated; leaves
    acuminate · · · · · · · · · · · · · · · · · · · · · · · · · · · · · · ·    12. *E. macilenta*

1. **E. kwaiensis** (*Engl.*) *Polh. & Wiens* in Lebrun & Stork, Énum. Pl. Fl. Afr. Trop.
2: 166 (1992) & Mistletoes Afr.: 123 (1998). Type: Tanzania, Lushoto District, W.
Usambara Mts., between Kwai and Gare, *Engler* 2230 (B!, holo., K!, fragment)

Stems spreading to 1 m. or so, whole plant glabrous; twigs slightly compressed, soon
thickened. Petiole 1–6 mm. long; lamina coriaceous, dark green above, paler beneath,
lanceolate to ovate-elliptic, 7–14 cm. long, 2–6 cm. wide, acuminate at the apex,
cuneate at the base, slightly sinuate at edges, with 5–8 pairs of lateral nerves. Flowers
2–3 per umbel in each axil and crowded at older thickened nodes; peduncle obsolete,
the bracts arising from cupular sockets formed by the pedicel; bract cupular, 1.5–2
mm. long. Receptacle 1.2–2 mm. long; calyx tubular, 2–3 mm. long, erose. Corolla
3.8–5 cm. long, tube orange to red, more yellow over vents, lobe-tips darker orange to
red; buds narrow, terminal swelling slight, 4 mm. long; tube slender, slightly
broadened upwards, 2.2–3.5 cm. long, with a short V-slit; lobes erect, linear-spathulate,
slightly hardened inside. Filaments transversely corrugate; tooth subulate to 1.5 mm.
long or often reduced to a small ledge; anthers 2.5–3 mm. long. Berry not seen.

TANZANIA. Lushoto District: Shume-Magamba Forest, 9 July 1983, *Polhill & Lovett* 4969! &
    Mkusi Forest near the road to Magamba, 19 Aug. 1952, *G.R. Williams* 504! & Mkusi [Mkuzu]
    Forest Reserve, on Magamba–Gare road, 2 Apr. 1969, *Ngoundai* 269!
DISTR. **T** 3; Malawi (Mt. Mulanje)
HAB. Rain-forest, probably on various hosts, but when on *Dasylepis integra* showing considerable
    similarity to leaf-form of the host; 1200–1800 m. (possibly lower in the E. Usambara Mts.)

SYN. *Loranthus kwaiensis* Engl. in E.J. 40: 522 (1908); Sprague in F.T.A. 6(1): 393 (1911); F.D.O.-
    A. 2: 178 (1932); T.T.C.L.: 285 (1949)
    *Tapinanthus kwaiensis* (Engl.) Danser in Verh. K. Akad. Wet., sect. 2, 29(6): 114 (1933)

NOTE. *E. kwaiensis* is still not very well known. Recent collections are all from the Shume-
Magamba Forest and Mazumbai area in the W. Usambara Mts., but Peter (1932) records it
from Monga in the E. Usambara Mts. and there is an unnumbered and imprecisely localised
specimen collected by Zimmermann in the early part of the century, presumably from near
Amani. The species has been collected rarely in submontane forest on Mt. Mulanje in
southern Malawi.
    *E. kwaiensis* and *E. longiflora* are most similar to a number of species in the forests of West
Africa. The reduced inflorescences of *E. kwaiensis*, with the flowers sessile in sockets formed
by the peduncle, and the red and yellow flowers are reminiscent of species in *Agelanthus*
sect. *Purpureiflori*, such as *A. sansibarensis* and *A. elegantulus*, but the stamens and style are
quite different.

2. **E. longiflora** *Polh. & Wiens*, Mistletoes Afr.: 123 (1998). Type: Tanzania, Iringa
District, Uzungwa Mts., Sanje, *J. Lovett* 297 (K!, holo., MO!, iso.)

Glabrous shrub to 1 m. or so; twigs slightly compressed. Petiole 2–4 mm. long;
lamina coriaceous, dark green, elliptic-lanceolate, 6–13 cm. long, 1–6 cm. wide,
acuminate at the apex, cuneate to slightly rounded at the base, sometimes slightly
undulate at edges, with 6–8 pairs of obscure lateral nerves. Flowers 2–4 in axillary
umbels; peduncles 1–2 mm. long; pedicels 1.5–6 mm. long; bract saucer-shaped with
a small limb, 1–2 mm. long, gibbous. Receptacle 1–2 mm. long; calyx cupular, 0.8–1
mm. long, ciliolate. Corolla 4–5 cm. long, red or orange to base of vents, yellow
around slits, darker orange or red above; buds slender, with a slight terminal swelling
3.5–6 mm. long; tube rather slender, slightly broadened upwards, 2.5–3.5 cm. long,

with a short V-split; lobes erect, linear-spathulate, very slightly hardened inside. Filaments green, transversely corrugated; tooth subulate, 0.7 mm. long; anthers 3 mm. long. Berry not seen.

TANZANIA. Morogoro District: Lusunguru Forest Reserve, *Mgaza* 295!; Ulanga District: Uzungwa Mts., Uhafiwa, 15 May 1991, *P. Lovett & Kayombo* 316!; Iringa District: Uzungwa Mts., Sanje, 29 Nov. 1983, *J. Lovett* 229!
DISTR. **T** 6, 7; known only from the Nguru and Uzungwa Mts.
HAB. Evergreen forest, found on *Psychotria*, *Rawsonia* and *Lasianthus*; 1150–1600 m.

3. **E. schubotziana** (*Engl. & K. Krause*) *Polh. & Wiens* in Lebrun & Stork, Énum. Pl. Fl. Afr. Trop. 2: 167 (1992); Balle in Toupin, Fl. Rwanda 1: 184, fig. 37/1 (1978), *nom. invalid.*, & in Troupin, Fl. Pl. Lign. Rwanda: 373, fig. 131/1 (1982), *nom. invalid.*; Polh. & Wiens, Mistletoes Afr.: 125, photo. 42 (1998). Type: Rwanda, Bugoye, *Mildbraed* 1436 (B!, holo., K!, fragment)

Twigs slightly compressed, usually scurfy puberulous with some short spreading hairs, glabrescent. Petiole 1–6 mm. long, shallowly channelled and hairy in the groove; lamina thin, yellow-green, linear-lanceolate to lanceolate, 3–12 cm. long, 0.8–3.5 cm. wide, acuminate at the apex, cuneate at the base, glabrous, with 5–8 pairs of lateral nerves, the lower 3–4 strongly ascending. Flowers 4–12 per umbel; peduncles 1–4 mm. long; pedicels 4–5 mm. long, hairy; bract triangular-ovate from a shallow cup, 2–3 mm. long, gibbous, ciliolate, puberulous. Receptacle 1.5–2.5 mm. long; calyx (1–)1.5–4 mm. long, ciliolate. Corolla 1.8–2.6 cm. long, orange or yellow-orange; buds 4-angled, isodiametric (unless insect galled) below, with a slight fusiform apical swelling 4–5 mm. long; tube 9–13 mm. long, with V-slit extending nearly to base, papillate inside, especially along sutures; lobes erect to spreading, perhaps sporadically reflexed, linear-spathulate, scarcely longer than tube. Stamens green, sometimes darkening with age; filament-tooth 0.7–1(–1.3) mm. long; anthers 3–4 mm. long. Berry not seen.

UGANDA. Toro District: Ruwenzori, Karangora, 27 Jan. 1935, *G. Taylor* 3262!; Ankole District: Rwampara, Rwenyaga Hill, Jan. 1940, *Eggeling* 3844!; Masaka District: Lake Nabugabo, *Lye* 2778!
TANZANIA. Bukoba District: Minziro, 5 Nov. 1994, *Congdon* 377!; Ngara District: Keza, Buseke, 20 May 1960, *Tanner* 4936!
DISTR. **U** 2, 4; **T** 1; E. Zaire and Rwanda
HAB. Forest edges and along rivers, on various hosts; 1150–2350 m.

SYN. *Loranthus schubotzianus* Engl. & K. Krause in E.J. 43: 311 (1909) & in Z.A.E. 2: 196 (1911); Sprague in F.T.A. 6(1): 387 (1911); Balle in F.C.B. 1: 325 (1948)
      *L. luteoaurantiacus* De Wild. in Rev. Zool. Afr. 9: B77 (1921) & Pl. Bequaert. 1: 318 (1922); Balle in F.C.B. 1: 324 (1948); F.P.N.A. 1: 97 (1948). Lectotype, chosen by Polh. & Wiens (1998): Zaire, Kivu, Ruwenzori, Butagu valley, *Bequaert* 3606 (BR!, lecto., BM!, isolecto.)
      *Tapinanthus luteoaurantiacus* (De Wild.) Danser in Verh. K. Akad. Wet., sect. 2, 29(6): 115 (1933)
      *T. schubotzianus* (Engl. & K. Krause) Danser in Verh. K. Akad. Wet., sect. 2, 29(6): 120 (1933)

NOTE. Closely related to the common species *E. woodfordioides*, but easily spotted in the field by the flowers not so obviously flushed blackish inside (the green stamens may be dark at least at the base).

4. **E. woodfordioides** (*Schweinf.*) *M.G. Gilbert* in Nordic Journ. Bot. 5: 223 (1985); Balle in Senck. Biol. 39: 108 (1958), *nom. invalid.*, & in Troupin, Fl. Rwanda 1: 184, fig. 37/2 (1978), *nom. invalid.*, & in Troupin, Fl. Pl. Lign. Rwanda: 373, fig. 131/2 (1982), *nom. invalid.*; Blundell, Wild Fl. E. Afr.: 129, t. 512 (1987); M.G. Gilbert in Fl. Ethiopia 3: 371, fig. 114.12 (1990); Polh. & Wiens in U.K.W.F., ed. 2: 155, t. 55 (1994) & Mistletoes Afr.: 126, photo. 43A–B. Type: Kenya, Nakuru/Laikipia District, Ewaso Narok [Guass Narok], *von Hoehnel* 39 (B!, holo.)

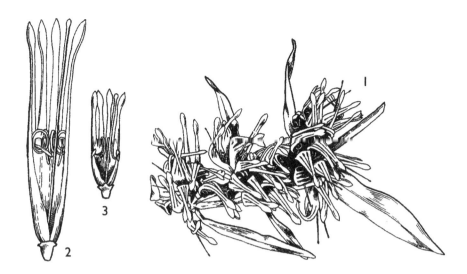

FIG. 7. *ENGLERINA WOODFORDIOIDES* – 1, flowering branch, × ¹/₅; 2, 3, flowers showing variation in size, × 2 ¹/₂. 1, from photo. of *M.G. & S. Gilbert* 1765; 2, from *Friis et al.* 467; 3, from *W. de Wilde* 6305. Drawn by Eleanor Catherine. Reproduced from Flora of Ethiopia

Shrub with spreading to drooping stems to 1 m. or so; twigs slightly compressed, glabrous. Petiole 3–10(–15) mm. long, grooved, the sides often crispate, the channel sometimes scurfy puberulous; lamina thin, coppery reddish when young, becoming fresh green, often with red veins, and a little more coriaceous, linear-lanceolate to oblong-lanceolate or ovate-lanceolate, 3–13(–15) cm. long, 0.6–3(–5) cm. wide, often attenuate, acuminate at the apex, base cuneate, rounded or rarely slightly cordate, glabrous except sometimes at youngest unfolded stage, with 4–8 pairs of lateral nerves, the lower 2–4 often (but not always) strongly ascending. Flowers 6–16(–20) per umbel; peduncle 1–5(–6) mm. long; pedicels (1–)2–5 mm. long; bract ovate-triangular from a shallow cup, 1–2 mm. long, slightly gibbous, ciliolate. Receptacle 1–1.5 mm. long; calyx 0.5–1 mm. long, ciliolate. Corolla (1.1–)1.5–2.5 cm. long, blue-grey with red lobes, inside tube blue-black; buds slightly inflated in lower part, narrow above vents to oblong-ellipsoid apical swelling (2–)2.5–3.5(–4) mm. long; tube (4–)6–12 mm. long, with V-slit to base, papillate inside especially along sutures; lobes erect, linear with spathulate tips, (7–)8–15 mm. long. Filaments blackish below, green above, tightly coiled; tooth 0.5–1.2 mm. long; anthers 1.5–3 mm. long. Berry bright red, obovoid, with prominent lobed disc, 6–9 mm. long, 3.5–5 mm. in diameter; seed orange. Fig. 7.

UGANDA. Karamoja District: Mt. Kadam [Debasien], 31 May 1939, *A.S. Thomas* 2957!; Kigezi District: Ishasha Gorge, Feb. 1950, *Purseglove* 3306!; Mbale District: Elgon, Bulago, 18 Aug. 1927, *Snowden* 1194!
KENYA. Northern Frontier Province: 12 km. N. of Maralal on road to Baragoi, 2 June 1979, *M.G. Gilbert, Kanuri & Mungai* 5442!; Nakuru District: 18 km. SW. of Molo on Keringet road, 7 Feb. 1985, *Kirkup* 31!; Nairobi City Park, 7 Aug. 1965, *Kokwaro* 240!
TANZANIA. Masai District: Loliondo, near District Officer's house, 6 July 1956, *G.R. Williams* 707!; Arusha National Park, Momella Gate, 17 Apr. 1968, *Greenway & Kanuri* 13478!; Moshi District: W. Kilimanjaro, Rongai, 6 Dec. 1968, *Shabani* 230!
DISTR. U 1–3; K 1–6; T 2, 3 (Pare Mts.); E. Zaire, Rwanda, Burundi and Ethiopia
HAB. Montane or riverine forest and associated bushland or wooded grassland nearby, on many hosts; 1350–3000 m.

SYN. *Loranthus woodfordioides* Schweinf. in L. Höhn., Rudolf- Stephanie-See: 856 (1892); Engl. in
E.J. 20: 126 (1894); Sprague in F.T.A. 6(1): 387 (1911); K. Krause in N.B.C.B. 8: 500
(1923); F.D.O.-A. 2: 176 (1932); Chiov., Racc. Bot. Miss. Consol. Kenya: 109 (1935); Balle
in F.C.B. 1: 321, t. 34 (1948); U.K.W.F.: 332, fig. on 331 (1974)
L. *ehlersii* Schweinf. in L. Höhn., Rudolf- Stephanie-See: 856 (1892); Engl. in E.J. 20: 126
(1894) & P.O.A. C: 167 (1895); Sprague in F.T.A. 6(1): 388 (1911); F.D.O.-A. 2: 177 (1932);
T.T.C.L.: 284 (1949). Type: Tanzania, Moshi District, S. Kilimanjaro, *Ehlers* (B!, holo.)
*Ischnanthus ehlersii* (Schweinf.) Tiegh. in Bull. Soc. Bot. Fr. 42: 260 (1895)
*I. woodfordioides* (Schweinf.) Tiegh. in Bull. Soc. Bot. Fr. 42: 260 (1895)
*Loranthus bagshawei* Rendle in J.L.S. 37: 206 (1905); Sprague in F.T.A. 6(1): 391 (1911).
Type: Uganda, Kigezi District, Rukiga [Ruchigga], *Bagshawe* 401 (BM!, holo.)
L. *adolfi-friderici* Engl. & K. Krause in E.J. 43: 312 (1909) & in Z.A.E. 2: 197 (1911); Sprague
in F.T.A. 6(1): 391 (1911); F.D.O.-A. 2: 178 (1932). Types: Rwanda, Rugege Forest,
*Mildbraed* 875 & 902 (B!, syn., K, fragments!)
L. *rugegensis* Engl. & K. Krause in E.J. 43: 313 (1909) & in Z.A.E. 2: 197, t. 17G–J (1911);
Sprague in F.T.A. 6(1): 392 (1911); F.D.O.-A. 2: 178 (1932); F.P.N.A. 1: 98 (1948). Type:
Rwanda, Rugege Forest, Rukarara R., *Mildbraed* 911 (B!, holo., K!, fragment)
L. *viminalis* Engl. & K. Krause in E.J. 43: 314 (1909) & in Z.A.E. 2: 199, t. 17A–F (1911);
Sprague in F.T.A. 6(1): 392 (1911); F.D.O.-A. 2: 178 (1932). Type: Rwanda, Rugege
Forest, *Mildbraed* 1043 (B,holo.!)
L. *umbelliflorus* De Wild. in Rev. Zool. Afr. 9: B81 (1921) & Pl. Bequaert. 1: 324 (1922).
Type: Zaire, Kivu, Mukule, *Bequaert* 5935 (BR!, holo.)
L. *brachyphyllus* Peter, F.D.O.-A. 2: 160, 178 & Descr. 13, t. 18/3 (1932); T.T.C.L.: 284
(1949). Type: Tanzania, S. Pare Mts., Shengena to Tona, *Peter* 9063 (B, holo.)
L. *longifolius* Peter, F.D.O.-A. 2: 160, 178 & Descr. 14, t. 19/2a (1932); T.T.C.L.: 284 (1949).
Type: Tanzania, S. Pare Mts., Shengena, *Peter* 8994 (B!, holo.)
L. *eucalyptoides* Peter, F.D.O.-A. 2: 160, 178 & Descr. 16, t. 20/2 (1932); T.T.C.L.: 284
(1949), *nom. illegit.*, *non* DC. (1840). Type: Tanzania, Arusha District, Mt. Meru, *Peter*
1888 (B!, holo.)
*Tapinanthus adolfi-friderici* (Engl. & K. Krause) Danser in Verh. K. Akad. Wet., sect. 2, 29(6):
107 (1933)
T. *bagshawei* (Rendle) Danser in Verh. K. Akad. Wet., sect. 2, 29(6): 108 (1933)
T. *ehlersii* (Schweinf.) Danser in Verh. K. Akad. Wet., sect. 2, 29(6): 111 (1933)
T. *rugegensis* (Engl. & K. Krause) Danser in Verh. K. Akad. Wet., sect. 2, 29(6): 119 (1933)
T. *umbelliflorus* (De Wild.) Danser in Verh. K. Akad. Wet., sect. 2, 29(6): 121 (1933)
T. *viminalis* (Engl. & K. Krause) Danser in Verh. K. Akad. Wet., sect. 2, 29(6): 121 (1933)
T. *woodfordioides* (Schweinf.) Danser in Verh. K. Akad. Wet., sect. 2, 29(6): 122 (1933)
T. *brachyphyllus* (Peter) Danser in Rec. Trav. Bot. Néerl. 31: 223 (1934)
T. *eucalyptoides* Danser in Rec. Trav. Bot. Néerl. 31: 223 (1934). Type as for L. *eucalyptoides* Peter
T. *longifolius* (Peter) Danser in Rec. Trav. Bot. Néerl. 31: 224 (1934)
*Loranthus woodfordioides* Schweinf. var. *minor* Chiov., Racc. Bot. Miss. Consol. Kenya: 109
(1935). Types: Kenya, Mt. Aberdare, Colle Roasia, *Balbo* 86 & 21 & Mt. Kenya, Nyeri
Forest, *Balbo* 58 & Karema, *Balbo* 44 (FT, syn.)
*Englerina woodfordioides* (Schweinf.) M.G. Gilbert var. *adolfi-friderici* (Engl. & K. Krause)
Balle in Senck. Biol. 39: 108, fig. 8–11 (1958) & in Troupin, Fl. Rwanda 1: 184 (1978) &
in Troupin, Fl. Pl. Lign. Rwanda: 373 (1982), *nom. invalid.*, *comb. non rite publ.*
E. *woodfordioides* (Schweinf.) M.G. Gilbert var. *umbelliflora* (De Wild.) Balle in Troupin, Fl.
Rwanda 1: 184 (1978) & in Troupin, Fl. Pl. Lign. Rwanda: 374 (1982), *nom. invalid.*,
*comb. non rite publ.*

NOTE. Easily recognised by the grey corolla-tube, which is blue-black within, and generally with
red lobes. Leaf-shape, venation and flower-size vary considerably and account for a number
of synonyms listed above. Much of the variation is sporadic and evidently trivial when seen in
the field, but some significant local divergence also occurs. Populations in SW. Ethiopia, at
the edge of the range from Jimma to Bonga, at relatively low altitudes (1700–2000 m.) and
relatively high rainfall, have small corollas (12–15 mm. long) with pale pinkish or apparently
even yellow lobes. A comparable small flowered form (*L. bagshawei*) is common in the W. Rift
populations (the species does not occur on Ruwenzori), but not clearly segregated from the
large-flowered forms, tending to predominate at lower altitudes but not exclusively so. By
contrast another small-flowered form occurs on the rain-shadow side of Mt. Kenya and
nearby parts of the eastern Aberdares and seems to be an independent derivative. The
populations from Meru and Kilimanjaro (*L. ehlersii*) are notably luxuriant, broad leaved and
floriferous, but again these populations at the edge of the range cannot be usefully or
precisely distinguished from the overall variation in the species.

5. **E. inaequilatera** (*Engl.*) *Gilli* in Ann. Naturhist. Mus. Wien 74: 422 (1971); Polh. & Wiens, Mistletoes Afr.: 127, photo. 14, 44 (1998). Lectotype, chosen by Polh. & Wiens (1998): Tanzania, Morogoro District, Uluguru Mts., Kifuru, *Stuhlmann* 9070 (B!, lecto., K!, fragment)

Shrub with spreading stems to 1 m. or so; twigs slightly compressed, scurfy pubescent with some short spreading hairs to glabrous. Petiole 1–7(–20) mm. long, shallowly channelled, often scurfy pubescent at least in the groove; lamina thin, dull pale green, often coppery reddish when young, linear-lanceolate to lanceolate or ovate, 3–12 cm. long, 0.5–4.5 cm. wide, acuminate at the apex, cuneate to rounded or rarely slightly cordate at the base, glabrous, with 6–10 pairs of lateral nerves, all spreading or the lower few more ascending. Flowers 4–12 per umbel; peduncle 1–4(–5) mm. long; pedicels 4–6(–8) mm. long, occasionally puberulous; bract ovate-triangular from a shallow cup, 1.5–2 mm. long, gibbous, ciliolate and sometimes puberulous. Receptacle 0.5–1.5 mm. long; calyx 0.5–1 mm. long, ciliolate. Corolla 2.5–3.5(–4) cm. long, yellow, orange or red, often more red on head, dark red inside tube; buds a little inflated in lower part, tapered to vents, then narrow to a slight oblong-ellipsoid apical swelling 3–5 mm. long; tube 1.1–1.6 cm. long, with V-slit often extending to base, papillate inside; lobes erect, linear-spathulate, 1.4–2 cm. long. Filaments deep maroon to blue-black, green above, tapered downwards into corolla-tube; tooth 1–2 mm. long; anthers 2.5–3 mm. long. Berry red to red-purple, sometimes partly white, obovoid, 6–7 mm. long, 4–5 mm. in diameter.

TANZANIA. Ufipa District: Mbisi Forest, 13 Mar. 1957, *Richards* 8700!; Morogoro District: Uluguru Mts., Lukwangule Plateau, above Chenzema Mission, 13 Mar. 1953, *Drummond & Hemsley* 1557!; Njombe District: 15 km. S. of Njombe, 5 July 1956, *Milne-Redhead & Taylor* 10956!
DISTR. T 4–8; easternmost Zambia, Malawi and Mozambique
HAB. Montane forest, often abundant at edges, extending down to drier forest on Rondo Plateau and into forest-woodland transition zones, on a wide variety of hosts; (700–)1400–2650 m.
SYN. *Loranthus inaequilaterus* Engl. in E.J. 28: 384 (1900); Sprague in F.T.A. 6(1): 390 (1911); F.D.O.-A. 2: 177 (1932); T.T.C.L.: 285 (1949)
    *L. lukwangulensis* Engl. in E.J. 28: 383 (1900); Sprague in F.T.A. 6(1): 388 (1911); F.D.O.-A. 2: 177 (1932); T.T.C.L.: 285 (1949). Type: Tanzania, Morogoro District, Uluguru Mts., Lukwangule Plateau, *Goetze* 301 (B!, holo., K!, fragment)
    *L. tenuifolius* Engl. in E.J. 30: 302 (1901); Sprague in F.T.A. 6(1): 389 (1911); Engl. in V.E. 3(1): 100, fig. 64 (1915); F.D.O.-A. 2: 177 (1932); Engl. & K. Krause in E. & P. Pf. 16B: 164, fig. 82 (1935); T.T.C.L.: 285 (1949), *nom. illegit., non* Tiegh. (1895). Type: Tanzania, Njombe District, Livingstone Mts., Yawiri Mt., *Goetze* 1194 (B!, holo., BR!, P!, iso., K!, fragment)
    *Tapinanthus inaequilaterus* (Engl.) Danser in Verh. K. Akad. Wet., sect. 2, 29(6): 113 (1933), as '*T. inaequilater*'
    *T. lukwangulensis* (Engl.) Danser in Verh. K. Akad. Wet., sect. 2, 29(6): 115 (1933)
    *T. tenuifolius* Danser in Verh. K. Akad. Wet., sect. 2, 29(6): 120 (1933). Type as for *L. tenuifolius* Engl.
    *Englerina tenuifolia* Gilli in Ann. Nat. Mus. Wien 74: 422 (1971). Type as for *L. tenuifolius* Engl.
NOTE. In the field it is readily apparent that this species complex is in the process of segregating into several taxa. The divergence is linked in part with the habitat, from forest through surrounding evergreen bushland to the transition zone of *Protea, Cussonia, Brachystegia* above the deciduous woodlands, and in part to a more precise exploitation of a particular suite of hosts in the more specialised habitats. This is generally accompanied by flowering flushes at different times of the year. In the Mufindi region a differential utilisation by Lycaenid butterflies has also developed (Congdon, pers. comm.). Within restricted areas these populations can be recognised by a suite of characters, but within the polymorphism of the whole species it seems impracticable to make any formal taxonomic distinctions yet.
    In the Mufindi area of the Southern Highlands, the common form is found throughout the forests, but is replaced by two distinctive variants north and westwards from Ifupira to Mafinga [Sao Hill]. The first is common on *Dodonaea* (an unusual host for mistletoes), but also on *Buddleja, Rhamnus, Jasminum* and *Protea,* in bushland and the forest-woodland

transition zone. The plants are small and bushy, with short flowering branches, the leaves small, yellow-green to mid-green, slightly coriaceous, lanceolate-elliptic, 1.5–4(–5.5) cm. long, the twigs persistently tomentellous, the fruits green (sometimes with longitudinal streaks) or white before ripening scarlet. The main flowering flush is around July–August at Ifupira. There is a further distinctive form at Mafinga, at the lower edge of the species range, in which the ovate leaves are notably coriaceous, dull mid-green and thick with age, and the berries are conspicuously yellow-green with darker longitudinal streaks before ripening scarlet. The growth form is more open than in the previous variant and the twigs tend to have a softer more floccose indumentum.

Plants from the Ukaguru and Nguru Mts. growing on *Vernonia* fall outside the normal range of variation, having exceptionally long petioles (8–20 mm.) and leaf-blades slightly cordate at the base, somewhat simulating the host.

6. **E. muerensis** (*Engl.*) *Polh. & Wiens* in Lebrun & Stork, Énum. Pl. Fl. Afr. Trop. 2: 166 (1992) & Mistletoes Afr.: 129 (1998). Lectotype, chosen by Sprague (1911): Tanzania, Lindi District, Rondo [Muera] Plateau, *Busse* 2868 (B!, lecto., BM!, BR!, EA!, isolecto., K!, fragment)

Shrub with spreading to pendulous branches; twigs slightly compressed, brownish tomentellous, glabrescent. Petiole 3–20 mm. long, puberulous to tomentellous; lamina thinly coriaceous, green, linear-lanceolate to narrowly elliptic-oblong or ovate-lanceolate, 4–16 cm. long, 0.8–4 cm. wide, slightly acuminate at the apex, cuneate at the base, with scurfy pubescence on midrib beneath and towards the base, otherwise subglabrous, with 6–8 pairs of lateral nerves, the lower 2–4 pairs ascending and looped some way from margin. Flowers 6–10 per umbel; peduncle, pedicels and bract tomentellous with white and brown hairs; peduncle 1–3 mm. long; pedicels 3–8 mm. long; bract ovate-triangular from a shallow cup, 2 mm. long, gibbous. Receptacle 1.5–2 mm. long; calyx 1–1.5 mm. long, ciliolate. Corolla 2.2–2.8 cm. long, orange, slightly yellowish towards the base, yellow inside tube, glabrous or sparsely pubescent especially along sutures; buds quadrangular and becoming slightly inflated in lower part, narrow above vents, with an oblong-ellipsoid apical swelling 4–5 mm. long; tube 1.1–1.3 cm. long, with V-slit extending to the base, papillate inside; lobes erect, linear-spathulate. Filaments deep green with purple-brown markings; tooth 1.2–1.5 mm. long; anthers 2–3 mm. long. Berry vermilion, obovoid, 6 mm. long, 5 mm. in diameter.

TANZANIA. Kilosa District: Mkata Juu, 10 July 1973, *Greenway & Kanuri* 15402!; Uzaramo District: Kazimzumbwi [Mzumbwi], 23 July 1939, *Vaughan* 2853!; Songea District: 1.5 km. W. of Tunduru District boundary, 6 June 1956, *Milne-Redhead & Taylor* 10640!
DISTR. **T** 6, 8; Malawi and N. Mozambique
HAB. *Brachystegia* and mixed woodland, on *Brachystegia, Combretum* or other Loranthaceae; 350–900 m.

SYN. *Loranthus muerensis* Engl. in E.J. 40: 529 (1908); Sprague in F.T.A. 6(1): 389 (1911); F.D.O.-A. 2: 177 (1932), excl. specim. ex Usambara; T.T.C.L.: 285 (1949), pro parte
      *Tapinanthus muerensis* (Engl.) Danser in Verh. K. Akad. Wet., sect. 2, 29(6): 116 (1933)

NOTE. Closely related to *E. inaequilatera*, differing by the conspicuous rusty tomentum on the twigs and inflorescence-axes and paler colours of the corolla and stamens. Generally found flowering in the dry season.

7. **E. cordata** *Polh. & Wiens* in Novon 7: 276 (1997) & Mistletoes Afr.: 129 (1998). Type: Malawi, Nyika National Park, Chisanga Falls, *Feehan* 100 (K!, holo., BR!, DBN!, K!, LISC!, MAL!, MO!, SRGH!, iso.)

Small glabrous shrub with mostly rather short spreading flowering branchlets; twigs slightly compressed. Leaves sessile, thin, glaucous, purple-flushed above when young, mostly lanceolate to ovate, 3–5 cm. long, 1–2 cm. wide, acuminate at the apex, basally cordate, with ± 6 pairs of lateral nerves (lower ones slightly more ascending), but leaves towards the base of flowering branchlets progressively smaller, blunter and rounded to oblate. Flowers 4–6 per umbel; peduncle 2–3 mm. long; pedicels 3–4 mm. long; bract acuminately triangular from a shallow cup, 2–3 mm. long, ciliolate at tip. Receptacle 1–1.5 mm. long, glaucous; calyx 0.3–0.4 mm. long, minutely ciliolate. Corolla-buds 2.5–2.7 cm. long, reddish with darker tip, a little swollen in lower part, tapered to vents 11 mm. from the base, then narrow to a slight oblong-ellipsoid apical swelling 5–6 mm. long, slightly apiculate; tube with very small scattered papillae inside, most numerous basally and along sutures; lobes erect, linear-spathulate. Filaments 10–11 mm. long; tooth subulate, 1.8 mm. long; anthers 3 mm. long. Berry white below, pink above, obovoid, 6 mm. long, 5 mm. in diameter; seed white.

TANZANIA. Ufipa District: Mbisi Forest Reserve, Nsonta Hill, 17 Nov. 1987, *Ruffo & Kisena* 2790!; Njombe District: 33 km. SW. of Luponde Tea Estate on Njombe–Manda road, 7 Oct. 1988, *Wiens & Spurrier* 7006! & Mlangali–Lisitu, 15 Nov. 1987, *Mwasumbi, Magehema & D. Thomas* 13776!
DISTR. **T** 4, 7; N. Malawi
HAB. Montane and riverine forest, on various hosts; 1800–2200 m.

NOTE. Similar to *E. inaequilatera* apart from the sessile cordate leaves.

8. **E. triplinervia** (*Baker & Sprague*) *Polh. & Wiens* in Lebrun & Stork, Énum. Pl. Fl. Afr. Trop. 2: 167 (1992) & Mistletoes Afr.: 131 (1998). Lectotype, chosen by Polh. & Wiens (1998): Tanzania, Uzaramo District, Dar es Salaam, *Kirk* (K!, lecto.)

Small shrub to 45 cm. or so, with numerous rather short (7–20 cm.) straight obliquely spreading flowering branchlets; twigs slightly compressed, pale, glabrous, markedly lenticellate. Petiole 0–3 mm. long; lamina chartaceous, yellow-green, lanceolate to narrowly elliptic or ovate-lanceolate, 1.5–6.5 cm. long, 0.8–2.5 cm. wide, attenuate to slightly acuminate at the apex, cuneate at the base, glabrous, with 3–6 pairs of lateral nerves, the basal pair ascending to middle or above. Flowers 2–6(–8) per umbel; peduncle 0.5–2 mm. long; pedicels 1.5–3 mm. long; bract bluntly triangular from a shallow cup, 1–1.5 mm. long, ciliolate. Receptacle 0.8–1 mm. long; calyx 0.2–0.3 mm. long, ciliolate. Corolla 9–12 mm. long, yellow, orange towards tip, with blackish streaks at vents; buds slightly 4-angular, only slightly inflated towards the base and apex; tube 3–5 mm. long, with V-slit extending almost to base, smooth inside (minute papillae only on edges of sutures); lobes erect, linear, only slightly broadened upwards, 6–7 mm. long. Filaments red, tightly coiled; tooth 0.2–0.5 mm. long, inserted ± that distance below anther; anthers 2–3 mm. long. Berry red to pinkish purple, obovoid, 8 mm. long, 4 mm. in diameter.

TANZANIA. Bagomoyo District: Mbegani, Voigt's I., 20 Jan. 1998, *Congdon* 508!; Uzaramo District: Kunduchi, 21 Nov. 1970, *B.J. Harris & Tadros* 5398! & University of Dar es Salaam, east of Primary School, 13 Aug. 1973, *Mwasumbi & Mhoro* 11194!
DISTR. **T** 6; N. Mozambique
HAB. Coastal bushland; 0–100 m.

SYN. [*Loranthus holstii* sensu P.O.A. C: 167 (1895), pro parte quoad distr. marit., *non* Engl. (1894)]
    *L. triplinervius* Baker & Sprague in K.B. 1911: 181 (June 1911) & in F.T.A. 6(1): 385 (Oct. 1011); F.D.O.A. 2: 175 (1932); T.T.C.L.: 284 (1949)
    *L. brachyanthus* Peter, F.D.O.-A. 2: 177 & Descr. 15, t. 20/1 (1932); T.T.C.L.: 284 (1949). Type: Tanzania, Uzaramo District, Dar es Salaam to Bagamoyo, *Peter* 44752 (B, holo.)
    *Tapinanthus triplinervius* (Baker & Sprague) Danser in Verh. K. Akad. Wet., sect. 2, 29(6): 121 (1933)
    *T. brachyanthus* (Peter) Danser in Rec. Trav. Bot. Néerl. 31: 223 (1934)

NOTE. A very characteristic small-flowered species from around Dar es Salaam and Bagamoyo, otherwise only known from Mozambique around Nampula.

9. **E. holstii** (*Engl.*) *Tiegh.* in Bull. Soc. Bot. Fr. 42: 257 (1895); Polh. & Wiens, Mistletoes Afr.: 132, photo. 47, fig. 10C (1998). Types: Tanzania, Lushoto District, W. Usambara Mts., Mlalo, *Holst* 325 (B!, syn., P!, isosyn.) & 2424 (B!, syn., K!, P!, US!, isosyn.) & Kwa Mshusa, *Holst* 8934 (B!, syn., BM!, K!, P!, isosyn.)

Shrub with slender branches to 1 m. or so; twigs compressed, sparsely scurfy pubescent to brownish tomentellous. Petiole 4–10 mm. long, puberulous at least in groove; lamina thin, fresh green with red midrib or flush, ovate-lanceolate to ovate, 2.5–7 cm. long, 1–3.5 cm. wide, acuminate at the apex, cuneate to rounded or slightly cordate at the base, sometimes slightly crisped along margins, scurfy puberulous when young, soon glabrescent except often near the base, with 7–12 pairs of spreading lateral nerves. Flowers 8–16 per umbel; peduncle 4–6(–8) mm. long, sparsely to densely hairy; pedicels 4–8 mm. long; bract ovate-triangular from a shallow cup, 1.5–3 mm. long, gibbous, puberulous. Receptacle 1–2 mm. long; calyx 0.5–1.5 mm. long, ciliolate. Corolla 1.8–2.3 cm. long, pink to red on tube, darker at vents and inside, with white to cream lobes with red tips; buds slightly 4-angular, bossed at vents, slightly wrinkled above, apical swelling slight, 5–6 mm. long; tube 4–5 mm. long, with irregular deep V-slit, smooth inside; lobes recurved to revolute, linear, slightly auricled at the base. Stamen-filaments 11–15 mm. long, dark red, loosely coiled, slightly expanded at the base; tooth 0.1–0.2 mm. long, a little below anther; anthers 3.5–3.8 mm. long. Berry dark red, obovoid, 5 mm. long, 3 mm. in diameter. Fig. 8/1–5.

TANZANIA. Lushoto District: Shagai Forest, near Sunga, 17 May 1953, *Drummond & Hemsley* 2573! & Mkusi, 8 July 1950, *G.R. Williams* 33! & Kwesimu, 28 July 1962, *Semsei* 3484!
DISTR. T 3; known only from the W. Usambara Mts.
HAB. Montane forest, especially abundant at forest edges, on various hosts, becoming a pest of plantations and orchards; 1200–2100 m.

SYN. *Loranthus holstii* Engl. in E.J. 20: 126 (1894) & P.O.A. C: 167, t. 15A–D (1895), pro parte; Sprague in F.T.A. 6(1): 385 (1911); F.D.O.-A. 2: 176 (1932), pro parte quoad specim. ex Usambara tantum; T.T.C.L.: 284 (1949), pro parte
   *L. crispulomarginatus* Engl. in E.J. 40: 529 (1908); F.D.O.-A. 2: 177 (1932). Type: Tanzania, W. Usambara Mts., Lushoto [Wilhelmstal], *Keudel* in Herb. *Amani* 616[III] (B, holo.)
   *L. holstii* Engl. var. *angustifolius* Engl. in F.D.O.-A. 2: 159, 176 (1932); T.T.C.L.: 284 (1949), nomen. Based on Tanzania, W. Usambara Mts., Wuga, Shashui, *Engler* 1113 (B†)
   *Tapinanthus holstii* (Engl.) Danser in Verh. K. Akad. Wet., sect. 2, 29(6): 113 (1933).

10. **E. kagehensis** (*Engl.*) *Polh. & Wiens* in Lebrun & Stork, Énum. Pl. Fl. Afr. Trop. 2: 166 (1992) & in U.K.W.F., ed. 2: 156 (1994) & Mistletoes Afr.: 132 (1998). Type: Tanzania, Mwanza District, Kayenzi [Kagehi], *Fischer* 537 (B!, holo.)

Small shrub, with rather short (5–20(–30) cm.) spreading flowering branchlets; twigs subterete to slightly compressed, hispidulous with short and very short spreading hairs. Leaves smaller towards the base of flowering branchlets, grading to bud-scales; petioles 0–3(–5) mm. long; lamina thinly coriaceous, dark green, elliptic-lanceolate to ovate, 1–6 cm. long, 0.6–2.5(–3.5) cm. wide, obtuse to acute, often slightly acuminate, at the apex (smaller proximal ones broader and blunter), minutely puberulous towards the base especially near the midvein, with 4–6 pairs of lateral nerves, lower ones or all ascending or curved-ascending. Flowers 4–6 per umbel; peduncle and pedicels subglabrous to hispidulous; peduncle 2–4 mm. long; pedicels 3–5 mm. long; bract bluntly triangular from a shallow cup, 1–1.5 mm. long, ciliolate. Receptacle 1–1.5 mm. long; calyx 0.3–0.5 mm. long, erose, ciliolate. Corolla 2–2.5 cm. long, greenish white to white or grey with pink tips; buds swollen at the base, bossed over vents, narrow, 4-angular and wrinkled above, ending in a narrow apical swelling 5 mm. long, slightly apiculate; tube 4–5 mm. long, with V-slit extending ± to base, hispidulous-papillate along sutures in upper half; lobes reflexed

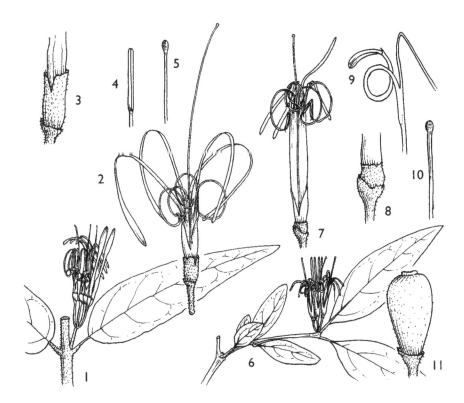

FIG. 8. *ENGLERINA HOLSTII* — **1**, flowering node, × 1; **2**, flower, × 3; **3**, base of flower, × 6; **4**, anther, × 4; **5**, style-tip, × 4. *ENGLERINA DRUMMONDII* — **6**, flowering branch, × 1; **7**, flower, × 3; **8**, base of flower × 4; **9**, stamen, showing attachment, × 4; **10**, style-tip, × 4; **11**, fruit, × 4. 1–5, from *Polhill & Lovett* 4987; 6, 11, from *Drummond & Hemsley* 2867; 7–10, from *Lovett* 248. Drawn by Christine Grey-Wilson.

to loosely revolute about the middle, linear-spathulate, expanded at the base with the inner face darker and hardened. Stamen-filaments 13–16 mm. long, dark red at the base, tapering upwards, slender, loosely coiled in upper part, extended at the base into a ridge ± 1 mm. down corolla-tube and minutely hispidulous-papillate or verruculose on the sides; tooth inserted a little below the anther, 0.4–0.7 mm. long; anthers 3–4.5 mm. long. Berry red, obovoid, 5 mm. long, 3–4 mm. in diameter; seed blue-grey.

KENYA. S. Kavirondo District: Usigu Market, 23 Feb. 1966, *Agnew et al.* 8051!; Masai District: Aitong, 1 Apr. 1961, *Glover, Gwynne & Samuel* 317!
TANZANIA. Musoma District: Kampi ya Pofu, 28 Feb. 1968, *Greenway & Kanuri* 13346!; Mpwapwa District: 11 km. S. of Gulwe on Kibakwe track, 9 Apr. 1988, *Bidgood, Mwasumbi & Vollesen* 966!; Iringa District: Nyangolo scarp, 62 km. N. of Iringa, 3 Feb. 1962, *Polhill & Paulo* 1333!
DISTR. **K** 5, 6; **T** 1, 4, 5, 7; not known elsewhere
HAB. Deciduous bushland, thicket, riverine fringes and wooded grassland (rarely in *Brachystegia* woodland), commonly on *Combretum, Grewia* or various legumes; 850–1750 m.

SYN. *Loranthus kagehensis* Engl. in E.J. 20: 129, t. 3G (1894) & P.O.A. C: 167 (1895); Sprague in F.T.A. 6(1): 386 (1911); F.D.O.-A. 2: 176 (1932); T.T.C.L.: 284 (1949)
    *Ischnanthus kagehensis* (Engl.) Tiegh. in Bull. Soc. Bot. Fr. 42: 260 (1895)
    *Tapinanthus kagehensis* (Engl.) Danser in Verh. K. Akad. Wet., sect. 2, 29(6): 113 (1933)
    *Loranthus sp. B* sensu U.K.W.F.: 332 (1974)

11. **E. ramulosa** (*Sprague*) *Polh. & Wiens* in Lebrun & Stork, Énum. Pl. Fl. Afr. Trop. 2. 167 (1992) & Mistletoes Afr.. 133 (1998). Types. Kenya coast, Mombasa–Takaungu, *Whyte* (K!, syn.) & Mombasa–Lamu, *Whyte* (K!, syn., B!, BM!, isosyn.)

Small shrub, with mostly short spreading flowering branches; twigs subterete to slightly quadrangular, with short spreading hairs, glabrescent, or sometimes almost glabrous. Leaves smaller towards the base of flowering branchlets, grading to bud-scales; petiole 0–1.5 mm. long, puberulous in groove; lamina thin, elliptic-lanceolate to ovate-elliptic (smallest ones ovate to circular), 1–3.5 cm. long, 0.7–1.8 cm. wide, obtuse at the apex, cuneate or rounded at the base, with small hairs on youngest, soon glabrescent, with 3–5 pairs of lateral nerves, lower 1–2 pairs strongly ascending. Flowers 2–6 per umbel; peduncle and pedicels 1.5–3 mm. long, glabrous or nearly so; bract ovate-triangular from a shallow cup, 1.5–2 mm. long, puberulous, ciliate. Receptacle 0.8–1 mm. long, as broad as tall; calyx 0.3–0.5 mm. long, ciliolate. Corolla 1.8–2.6 cm. long, yellow, with a touch of red at vents; buds 4-angular, slightly swollen at the base, bossed over vents, slightly wrinkled above, with a slight apical swelling 6–9 mm. long, apiculate; tube 4–6 mm. long, with V-slit extending nearly to base, minutely papillate along sutures; lobes reflexed to revolute, linear, slightly dilated at the base. Stamen-filaments orange, tightly coiled, dilated, with minute papillae at the very base; tooth 1 mm. long; anthers 5.5–6.5 mm. long. Berry obovoid, 6 mm. long, 4 mm. in diameter.

KENYA. Kilifi District: Malindi, June 1962, *Tweedie* 2386! & 16 June 1983, *S.A. Robertson* 3600!; Lamu District: 16 km. from Kiunga on Lamu road, 13 Aug. 1961, *Gillespie* 184!
DISTR. **K** 7; S. Somalia
HAB. Coastal bushland; 0–60 m.

SYN. *Loranthus ramulosus* Sprague in K.B. 1911: 182 (June 1911) & in F.T.A. 6(1): 386 (Oct. 1911); Chiov., Fl. Somala 2: 358, fig. 223 (1932)
*Tapinanthus ramulosus* (Sprague) Danser in Verh. K. Akad. Wet., sect. 2, 29(6): 118 (1933)

12. **E. macilenta** *Polh. & Wiens*, Mistletoes Afr.: 133 (1998). Type: Tanzania, Dodoma District, Kilimatindi, *Prittwitz* 132 (BM!, holo.)

Small well-rounded shrub to 0.5 m., with mostly rather short spreading flowering branchlets; twigs subterete, glabrous or with very short spreading glandular hairs. Leaves slightly smaller towards the base of flowering branchlets; petiole 2–3 mm. long, glandular-puberulous in the groove; lamina thinly coriaceous, narrowly elliptic, 2–4.5 cm. long, 0.6–1.5 cm. wide, acuminate at the apex, cuneate at the base, glabrous, with 4–8 sometimes obscure pairs of lateral nerves, lowermost more ascending. Flowers 2–3 per umbel; peduncle 1–2 mm. long; pedicels 2–2.5 mm. long; bract triangular from a shallow cup, 1.5–2 mm. long, ciliolate. Receptacle 1 mm. long; calyx 0.5 mm. long, ciliolate. Corolla 2–2.2 cm. long, white to yellow tipped red; buds parallel-sided and slightly corrugated to vents, slightly narrowed to base, narrow above vents to a slight apical swelling 6–7 mm. long; tube 6–7 mm. long, with a V-slit to the middle, the surface corrugated with a row of dimples either side of the filament-line, sometimes distinctly papillate along the sutures above; lobes reflexed to revolute, linear-spathulate, slightly dilated at the base. Stamen-filaments inrolled, dilated at the base, sometimes with minute papillae around basal thickening; tooth 0.5 mm. long and inserted that distance below anther; anthers 4–5 mm. long. Berry not seen.

TANZANIA. Mpwapwa District: Mangalisa to Ikuyu, 12 Apr. 1988, *Lovett & Congdon* 3240!; Iringa District: Madibira, 5 Mar. 1989, *Congdon* 222!; Lindi District: Kiwetu, 1 Apr. 1991, *Fison* 91/5!
DISTR. **T** 5, 7, 8; not known elsewhere
HAB. Mixed deciduous bushland; 600–1050 m.

SYN. [*Loranthus holstii* sensu F.D.O.-A. 2: 176 (1932), pro minore parte, quoad specim. *Prittwitz* 132; T.T.C.L.: 284 (1949), pro minore parte, *non* Engl. sensu stricto]

13. **E. drummondii** *Polh. & Wiens*, Mistletoes Afr.: 134, fig. 10D (1998). Type: Tanzania, Lushoto District, W. Usambara Mts., Mkuzi, *Drummond & Hemsley* 1337 (K!, holo., B!, SRGH!, iso.)

Small shrub with short spreading flowering branchlets mostly 5–20 cm. long; twigs subterete to slightly quadrangular, scurfy puberulous at least on younger internodes, soon glabrescent. Leaves smaller towards the base of flowering branchlets, grading to bud scales; petiole 0–4 mm. long, scurfy puberulous in groove; lamina thin, elliptic-lanceolate to ovate-elliptic, 2–6 cm. long, 0.8–2.5 cm. wide, acuminate at the apex, cuneate at the base (small proximal leaves broader and blunter), scurfy puberulous, soon glabrescent, with 3–6 pairs of lateral nerves, lower 1–3 pairs strongly ascending. Flowers 2–4 per umbel; peduncles 0.5–2 mm. long, subglabrous or with minute hairs; pedicels 2.5–5 mm. long; bract ovate-triangular from a cupular base, 1.5–2 mm. long, ciliolate. Receptacle 0.8–1 mm. long, broader than long; calyx 0.3–0.5 mm. long, ciliolate. Corolla (1.3–)1.7–2.2 cm. long, yellow-orange, sometimes more yellow or greenish at the base, with red tips to buds and expanded lobes; buds slightly 4-angular, slightly swollen at the base, slightly bossed over vents, with a slight apical swelling 4 mm. long, a little wrinkled below apical swelling; tube (5–)7–9 mm. long, with V-slit extending nearly to the base, smooth inside; lobes reflexed or revolute about the middle, linear-spathulate. Stamen-filaments dark red or brown, tightly coiled, extended at the base as a ridge 2–3 mm. down the corolla-tube; tooth 0.4–0.5 mm. long; anthers 2.5–3 mm. long. Berry bright red, sometimes white at the base, pear-shaped to almost obconic, 5–7 mm. long, 4 mm. in diameter. Fig. 8/6–11 (p. 35).

KENYA. Kitui District: 4 km. N. of Kangondi on Embu road, 16 Dec. 1971, *Wiens* 4484! & 19 Mar. 1985, *Kirkup* 42!; Machakos District: Mbuinzau [Mbinzao], 29 Jan. 1942, *Bally* 1741!; Teita District: Kasigau, 16 Nov. 1994, *Luke* 4133!
TANZANIA. Lushoto District: W. Usambara escarpment, Gologolo–Mkumbala footpath, 4 June 1953, *Drummond & Hemsley* 2867! & Mkuzi–Kifungilo, Mar. 1959, *Mgaza* 245! & below Baga II Forest Reserve, between Mgwashi and Mtai above Mzinga village, 31 Jan. 1985, *Borhidi, Iversen & Mziray* 85418!
DISTR. K 4, 7; T 3; not known elsewhere
HAB. Dry evergreen forest, hilltop mist forest and nearby wooded grasslands; 1000–1950 m.

SYN. *Loranthus sp. C* sensu U.K.W.F.: 332 (1974), pro parte quoad specim. ex "MAC"
*Englerina sp. A* sensu Polh. & Wiens in U.K.W.F., ed. 2: 156 (1994)

14. **E. heckmanniana** (*Engl.*) *Polh. & Wiens* in Lebrun & Stork, Énum. Pl. Fl. Afr. Trop. 2: 166 (1992); Blundell, Wild Fl. E. Afr.: 129, t. 469 (1987), *nom. invalid.*; Polh. & Wiens in U.K.W.F., ed. 2: 155 (1994) & Mistletoes Afr.: 135 (1998). Type: Tanzania, Iringa District, Mbijiri [Mbigiri], *Goetze* 509 (B!, holo., BR!, K!, iso.)

Small shrub with mostly rather short (5–15 cm.) spreading flowering branchlets; twigs slightly compressed, hispidulous to tomentellous with short and very short spreading hairs. Leaves smaller towards the base of flowering branchlets, grading to bud scales; petioles 0–1 mm. long; lamina thin, ovate to elliptic-oblong or elliptic-lanceolate, 1–4 cm. long, 0.8–2.5 cm. wide, obtuse to acute, scarcely acuminate, at the apex, broadly cuneate to rounded at the base, scabridulous-pubescent on both surfaces, with 4–8 pairs of lateral nerves, the lower ones more ascending. Flowers (1–)2–6 per umbel; peduncle and pedicels 1–2 mm. long, densely hispidulous, bract ovate from a very slight cup, 2–3 mm. long, densely hairy. Receptacle 1.2–1.5 mm. long; calyx 0.3–0.8 mm. long, ciliate. Corolla 3.2–4.5 cm. long, yellow or orange-yellow, often a little more orange or reddish at tip, green at the vents, ciliate at angles (sutures) to densely hairy overall; buds 4-angled, bossed at vents, scarcely swollen in upper part, pointed; tube 8–10 mm. long, with V-slit extending halfway, smooth inside except sometimes for minute papillae along sutures above; lobes reflexed to revolute about the middle, linear, slightly tapering to middle, dilated at the base. Stamen-filaments rather loosely coiled, dilated and thickened at the base, subcordate, sometimes minutely papillate around base; tooth 0.2–0.7 mm. long; anthers 7.5–9 mm. long. Berry not seen.

subsp. **heckmanniana**; Polh. & Wiens, Mistletoes Afr.: 136 (1998)

Corolla ciliate at angles, along edges of lobes and sutures of the tube, glabrous on the surfaces.

TANZANIA. Dodoma District: 80 km. on Dodoma–Iringa road, 18 Feb. 1932, *St. Clair-Thompson* 376!; Kilosa District: 3 km. N. of Tanzam highway on road to Malalo, E. side of Great Ruaha R., 25 Jan. 1986, *D. & C. Wiens* 6578c!; Iringa District: 56 km. from Iringa on Dodoma road, 26 Feb. 1961, *Verdcourt* 3081!
DISTR. **T** 5–7; known only from the Iringa Plateau, together with its scarps and fringes
HAB. Deciduous bushland and thicket, on various hosts including *Combretum* and *Grewia*; 950–1550 m.

SYN. *Loranthus heckmannianus* Engl. in E.J. 28: 384 (1900); Sprague in F.T.A. 6(1): 384 (1910); F.D.O.-A. 2: 175 (1932); T.T.C.L.: 285 (1949); U.K.W.F.: 330 (1974)
*Tapinanthus heckmannianus* (Engl.) Danser in Verh. K. Akad. Wet., sect. 2, 29(6): 113 (1933)

subsp. **polytricha** *Polh. & Wiens*, Mistletoes Afr.: 136, photo. 48 (1998). Type: Kenya, Machakos District, Nairobi–Mombasa road, 21 km. from Sultan Hamud, *Verdcourt* 3809 (K!, holo.)

Corolla hispidulous-tomentose overall; axes and leaves also more densely hairy than typical subspecies.

KENYA. Naivasha District: Gilgil, 12 Jan. 1932, *V.G. van Someren* in *C.M.* 1691!; Machakos District: Kilima Kiu Estate, 7 km. SE. of Nairobi–Mombasa highway, 14 Dec. 1971, *Wiens* 4480!; Masai District: 1.5 km. S. of Masai village on W. side of Ngong Hills, 29 Jan. 1985, *Kirkup* 21!
TANZANIA. Shinyanga, 25 Jan. 1934, *B.D. Burtt* 5057!; Masai District: Loliondo, Kingarana Forest Reserve, 18 Mar. 1995, *Congdon* 420!; Dodoma District: ± 10 km. NW. of Mtera bridge, 15 Aug. 1970, *Thulin & Mhoro* 753!
DISTR. **K** 3, 4, 6; **T** 1–3, 5; known only along the Great Rift Valley and its flanks from central Kenya to central Tanzania
HAB. Edges of upland dry evergreen forest and associated *Tarchonanthus* bushland or in deciduous bushland, on a variety of hosts including *Grewia*, *Combretum*, *Acacia*, *Cordia*, *Teclea*, *Commiphora* and *Euclea*; 700–2300 m.

NOTE. Abuts the range of subsp. *heckmanniana* at the foot of the Iringa scarp. *Thulin & Mhoro* 753, cited above, might be regarded as slightly intermediate.

## 8. AGELANTHUS

Tiegh. in Bull. Soc. Bot. Fr. 42: 246 (1895); Polh. & Wiens, Mistletoes Afr.: 137 (1998)

*Dentimetula* Tiegh. in Bull. Soc. Bot. Fr. 42: 265 (1895)

Shrubs, mostly 0.5–2 m. from a single haustorial connection, rarely with subcortical runners; twigs terete to slightly compressed or angular; hairs, if present, simple or irregularly branched. Leaves alternate to opposite, sometimes crowded on short shoots, shortly petiolate; lateral nerves spreading-ascending to strongly ascending from the base. Flowers borne in sessile to pedunculate heads or umbels, clustered or rarely singly in the axils, sometimes terminal on short shoots; bract shallowly to distinctly cupular, with a small to leafy limb. Calyx saucer-shaped to tubular, entire to slightly toothed. Corolla 5-lobed, the lobes relatively short, occasionally (*A. platyphyllus*) longer than the tube, generally conspicuously banded in different colours; mature buds developing vents below the tip (bud-apex swollen or not) and opening with a short V-split; basal swelling present or absent; lobes linear-lanceolate to linear-elliptic above the claw, generally remaining erect, sometimes shiny and hardened inside. Filaments inserted near top of corolla-tube or on claw if lobes exceptionally long, short, inflexed or rarely inrolled, sometimes corrugated, sometimes hardened at tip, with or without a ledge or tooth in front of the anther; anthers 4-thecous, linear, with connective not or only slightly produced at the apex. Style isodiametric or swollen opposite the filaments and constricted above (skittle-shaped); stigma small, capitate. Berry ellipsoid to obovoid, smooth to warty, usually ripening red.

57 species in Africa and the Arabian peninsula.

This is the largest genus in Africa and shows considerable modification in the structures of the flower related to pollination. The species have generally been grouped with or close to *Tapinanthus* (*Loranthus* sect. *Constrictiflori* (Engl.) Sprague), but are readily distinguished from the species included here in *Tapinanthus* by the colour-banded flowers with vents opening in the mature buds, the corolla-lobes generally remaining erect.

The six sections are all represented in East Africa.

Sect. 1. **Longiflori** (*Engl.*) *Polh. & Wiens* (syn. *Schimperina* Tiegh.; *Loranthus* sect. *Dendrophthoë* group *Longiflori* Engl.; *L.* sect. *Longiflori* (Engl.) Sprague). Bud-scales at base of shoots conspicuous, somewhat persistent. Leaves alternate to subopposite, pinnately nerved, lower ones on each branch sometimes smaller. Flowers in axillary pedunculate heads; bract saucer-shaped with a small triangular limb. Calyx saucer-shaped. Corolla-buds subcylindrical, apiculate, not swollen at base and only slightly swollen over anthers; lobes relatively long, sometimes longer than the tube, weakly erect, sometimes coherent. Filaments at the top of corolla-tube or arising from claws of lobes, not toothed, tip narrowed and hardened; anthers elongate. Style nearly isodiametric to slightly skittle-shaped. Berries smooth. Species 1–2.

Sect. 2. **Acranthemum** (*Tiegh.*) *Polh. & Wiens* (syn. *Acranthemum* Tiegh.; *Loranthus* sect. *Acranthemum* (Tiegh.) Sprague). Bud-scales at base of shoots conspicuous, subpersistent. Leaves opposite to alternate, partly crowded on short shoots, penninerved. Flowers terminal on short shoots or in umbels from clusters of leaves or axillary; bract small. Calyx saucer-shaped. Corolla-buds subcylindrical, apiculate, only slightly swollen at base, with subdued colour-banding; lobes longer than in following sections, not hardened. Filaments inflexed, tapered, not hardened, with or without a small ledge in front of the anther. Style isodiametric to skittle-shaped. Berry smooth. Species 3.

Sect. 3. **Obtectiflori** (*Engl.*) *Polh. & Wiens* (syn. *Loranthus* sect. *Tapinanthus* group *Obtectiflori* Engl.; *L.* sect. *Obtectiflori* (Engl.) Sprague). Shoots with small but evident bud-scales at base. Leaves opposite or subopposite, with spreading-ascending nerves, lower leaves on side branches smaller and more obovate. Flowers in pedunculate umbels in axils or less often terminal on short shoots; bract small to leafy. Calyx saucer-shaped. Corolla subcylindrical in bud, slightly apiculate, scarcely swollen at base, multicoloured (exceptionally almost entirely white); lobes not hardened. Filaments tapered, not hardened, with a small tooth. Style isodiametric. Berry smooth. Species 4–5.

Sect. 4. **Agelanthus**. Bud-scales inconspicuous. Leaves alternate to opposite, elsewhere sometimes crowded, usually with few strongly ascending nerves from base, sometimes ± pinnately nerved. Flowers clustered in axils or in very shortly pedunculate axillary umbels; bract cupular with a small limb. Calyx saucer-shaped to cupular or tubular. Corolla subcylindrical in bud or conspicuously swollen at base, usually with strong colour banding in various combinations; lobes not hardened. Filaments slightly tapered, sometimes corrugated with a hardened tip, generally without a tooth in front of the anther. Style isodiametric or slightly skittle-shaped, but then thickened part winged or cross-shaped in section. Berry smooth, rarely verruculose elsewhere. Species 6–13.

Sect. 5. **Purpureiflori** (*Engl.*) *Polh. & Wiens* (syn. *Dentimetula* Tiegh.; *Loranthus* sect. *Tapinanthus* group *Purpureiflori* Engl.; *L.* sect. *Purpureiflori* (Engl.) Sprague). Leaves with few strongly ascending nerves from base. Flowers in sessile or very shortly pedunculate umbels, with pedicels in sockets; bracts cupular. Calyx cupular to tubular. Corolla subcylindrical in bud or with a marked basal swelling, red banded yellow over vents, lobes glossy inside but not hardened. Filament-tips pale, hardened and thickened (rather like a crab's claw), with a distinct tooth in front of the anther. Style skittle-shaped. Berry usually warted. Species 14–22.

Sect. 6. **Erectilobi** (*Sprague*) *Polh. & Wiens* (syn. *Loranthus* sect. *Erectilobi* Sprague, pro majore parte). Leaves penninerved, with lower nerves ± strongly ascending. Flowers in shortly pedunculate umbels; bracts saucer-shaped with a small limb. Calyx saucer-shaped to shortly cupular. Corolla with ± swollen head in bud and markedly swollen base, yellow to red, banded over vents, reddish at tip of lobes; lobes usually glossy and hardened inside (least so in *A. molleri*). Filaments rarely hardened, toothed. Style skittle-shaped. Berry smooth. Species 23–35.

1. Corolla subcylindrical, without or with only a slight elongate swelling at the base, only slightly swollen over the anthers in bud · · · · · · · · · · · · · · · · · · · · · · · · · · · · · · · · · · · · · · 2
   Corolla conspicuously swollen at base, narrowly constricted above, with or without a somewhat swollen apex* · · · · · · · · · · · · · · · · · · · · · · · · · · · · · · · · · · · · · · · · · · · · 10
2. Inflorescences 2–4-flowered umbels terminal on short shoots with a cluster of small leaves (sect. *Acranthemum*)    3. *A. microphyllus*
   Inflorescences all or almost all axillary on long shoots, sometimes clustered at old nodes · · · · · · · · · · · · · · · · · · · · · · · · · · · 3
3. Flowers in heads or umbels on peduncles 4–40 mm. long, generally 8–24 and if fewer then peduncle especially well developed; leaves with spreading-ascending nerves · · · · · · · · · · · · · · · · 4
   Flowers in small umbels on peduncles 0–3 mm. long, mostly 2–6 per umbel (often several umbels clustered together at the nodes); leaves with 3–5 strong nerves from near base except in *A. dodoneifolius* and *A. tanganyikae* · · · · · · · · · · · · · · · · · · · · · · · · · · · · · · · · · · · · · 7
4. Corolla-lobes 10–27 mm. long, the upper part linear to linear-lanceolate, forming a narrowly fusiform and apiculate tip to the buds; filaments without a tooth in front of the anther; leaves all similar, grading from alternate to opposite (sect. *Longiflori*) · · · · · · · · · · · · · · · · · · · · · · · · · 5
   Corolla-lobes 8–12 mm. long, the upper part elliptic or oblong-elliptic, forming an ellipsoid and only slightly apiculate head to the mature buds; anthers with a small tooth in front of the anther; leaves tending to be smaller on shortened lateral flowering shoots (sect. *Obtectiflori*) · · · · · · · · · · · · · · · · · · · · · · · · · · · · · · · · · · · · · 6
5. Corolla-lobes 10–12 mm. long; anthers 3–3.5 mm. long; peduncle 4–8 mm. long; leaves mostly lanceolate · · ·    1. *A. unyorensis*
   Corolla-lobes 20–27 mm. long; anthers 7–8.5 mm. long; peduncle 10–18 mm. long; leaves usually ovate to elliptic · · · · · · · · · · · · · · · · · · · · · · · · · · · · ·    2. *A. platyphyllus*
6. Bracts with limb lanceolate to oblanceolate or leafy, 2–many times longer than the saucer-shaped basal part · · · · · · · · · · · · · · · · · · · · · · · · · · · · · · ·    4. *A. subulatus*
   Bracts small, the triangular-ovate limb not much longer than the basal part · · · · · · · · · · · · · · · · · · · · · · ·    5. *A. longipes*
7. Corolla 1.8–3.3 cm. long, pink to cream with lobes green to red above, the buds with slightly clavate heads · · ·    7. *A. kayseri*
   Corolla 4–7 cm. long, dull red, paler or greenish to yellow over vents, the buds with narrowly fusiform heads · · · · · · · · · · · · · · · · · · · · · · · · · · · · · · · · · · · · · 8
8. Pedicels 1–2 mm. long, tapered down from the bract; filaments slightly hardened, narrowed and flattened at tip, without a tooth; leaves shortly petiolate at first, but with lamina decurrent to base and expanding with age so leaves become more sessile · · · · · · · · · · · · · · · · · · · · · · · · · · · · · · 9

---

\* Species in sect. *Purpureiflori* (species 15–17) with a slight, elongate swelling are accommodated under both leads (hence repetition of same distinction at dichotomies 8 and 12). It should be noted, however, that the basal swelling of the corolla only develops as the buds mature. If keying material with only very young buds both leads will have to be explored to a greater extent, especially for species in sect. *Agelanthus* related to *A. brunneus* (species 8–13), as noted under that species.

Pedicels within obliquely cupular sockets (thus three cups formed by pedicel-socket, bract and calyx); filaments hardened and swollen at tip like a crab's claw (just flattened and hardened in *A. dodoneifolius*), with a small tooth in front of the anther; leaves tapered into a short distinct petiole · · · · · · · · · · · · · · · · · · · · · · · · · · · · · 17

9. Bract ovate-triangular from a shallowly cupular base, not obscuring whole receptacle; calyx cupular, 1–2 mm. long, usually not split · · · · · · · · · · · · · · · · · · · · · · ·   10. *A. djurensis*

Bract cupular, with scarcely any limb, obscuring the receptacle in flower; calyx deeply cupular, 3 mm. long, split by slightly expanding base of corolla · · · · · · · · ·   11. *A. toroensis*

10. Corolla 2.5 cm. long (or possibly a little longer), yellow at base, purple-red above, delicate, with lobes ± half as long as tube; flowers in shortly pedunculate umbels; leaves penninerved · · · · · · · · · · · · · · · · · · · · · · ·   6. *A. rondensis*

Corolla larger, more robust, differently coloured and with relatively short lobes · · · · · · · · · · · · · · · · · · · · · · · · · · · · 11

11. Flowers clustered in the axils or in sessile to very shortly pedunculate umbels; pedicels short, cup-shaped or lacking; corolla-lobes not hardened and glossy inside; leaves sometimes strongly 3(–5)-nerved from near the base · · · · · · · · · · · · · · · · · · · · · · · · · · · · · · · · · · · · · · · · · · · · · 12

Flowers in distinctly pedunculate umbels even if peduncle quite short and with distinct pedicels; corolla-lobes hardened and glossy inside (least so in *A. molleri*); leaves pinnately nerved, the lower nerves ± strongly ascending (sect. *Erectilobi*) · · · · · · · · · · · · · · · · · · · · · · · · 24

12. Pedicels lacking or tapered down from bract; filaments without a tooth in front of the anther (sect. *Agelanthus*) · · · · · · · · · · · · · ·13

Pedicels in cupular sockets, surrounding base of cupular bract; filaments with a tooth in front of the anther (sect. *Purpureiflori*) · · · · · · · · · · · · · · · · · · · · · · · · · · · · · · · · · · · · 16

13. Corolla pink to reddish, banded at top of tube, lobes white to yellowish over the vents in bud, and red at tips* · · · · · · · · · · · · · · · · · · · · · · · · · · · · · · · · · · · · · · · · · · 14

Corolla yellow or yellow-orange, similarly banded at top of tube, lobes white or pinkish over vents, tips red · · · · · · · · · · · · · · · · · 15

14. Corolla-lobes above claw linear-oblanceolate, broadest above middle; leaves green or yellowish green, persistently attenuate-cuneate at the base, so leaves become distinctly more petiolate with age; branchlets slightly flattened then angular, glabrous to hairy · · · · ·   8. *A. zizyphifolius*

Corolla-lobes above claw linear-elliptic, broadest about the middle; leaves usually light green to dark green, the lamina expanding on the attenuate part at the base so leaves become more sessile; branchlets angular but not flattened, glabrous · · · · · · · · · · · · · · · · · · · · ·   9. *A. brunneus*

15. Calyx 3–4 mm. long; lamina 3–4 times as long as broad, basally cuneate; anthers 1.8–2 mm. long · · · · · · · · · ·   12. *A. krausei*

Calyx 2–2.5 mm. long; lamina 1.5–3 times as long as broad, broadly cuneate to rounded at base; anthers 2–2.5 mm. long · · · · · · · · · · · · · · · · · · · · · · · · · · ·   13. *A. pennatulus*

* If corolla-tube dark red see species 10 and 11 at dichotomy 9.

16. Calyx campanulate to tubular, usually at least twice as
    long as broad · · · · · · · · · · · · · · · · · · · · · · · · · · · · · · · · · · · · · · · · · · · 17
    Calyx (excluding solid receptacle below) cupular, at least
    as broad as long · · · · · · · · · · · · · · · · · · · · · · · · · · · · · · · · · · · · · · · · 21
17. Corolla-lobes 12–17 mm. long · · · · · · · · · · · · · · · · · · · · · · · · · · · · · · · 18
    Corolla-lobes 7–10 mm. long · · · · · · · · · · · · · · · · · · · · · · · · · · · · · · · 20
18. Calyx split unilaterally by developing basal swelling of
    corolla-tube, leaves elongate, mostly 4–10 or more
    times as long as broad · · · · · · · · · · · · · · · · · · · · · · ·      14. *A. dodoneifolius*
    Calyx circumscissile, splitting around base, a tubular
    remnant remaining on corolla-tube above the basal
    swelling; leaves ovate-elliptic, 1.5–3 times as long as
    broad · · · · · · · · · · · · · · · · · · · · · · · · · · · · · · · · · · · · · · · · · · · · · · · · 19
19. Branchlets tomentellous; corolla-lobes 14–17 mm. long      16. *A. irangensis*
    Branchlets glabrous; corolla-lobes 12 mm. long · · · · · ·      17. *A. validus*
20. Corolla 4.5–6 cm. long, with a negligible to slight basal
    swelling not splitting the calyx or splitting it down one
    side; leaves with several pairs of ascending nerves · · ·      15. *A. tanganyikae*
    Corolla 3.2–4.2 cm. long, with a globular basal swelling
    that splits the calyx around its base, the remnant
    remaining as a tubular ring on the corolla; leaves with
    second pair of lateral nerves strongly ascending · · · ·      19. *A. bipartitus*
21. Leaves bluish green; umbels with peduncles 0–2(–3)
    mm. long bearing pedicel-sockets laterally, pedicels
    0–2 mm. long within the socket · · · · · · · · · · · · · · ·      22. *A. fuellebornii*
    Leaves green to yellow-green, sometimes red flushed;
    umbels sessile, pedicels not clearly differentiated · · · · · · · · · · · · · · · 22
22. Leaves oblanceolate to obovate, thick, 3–5-nerved from
    base · · · · · · · · · · · · · · · · · · · · · · · · · · · · · · · · · · · · ·      21. *A. oehleri*
    Leaves broadest about the middle, thinly fleshy, mostly
    with second to third pairs of nerves strongly ascending · · · · · · · · · · · · · · 23
23. Leaves mostly narrowing gradually toward tip; basal
    swelling of corolla 2.5–3 mm. across; calyx toothed,
    split unilaterally by expanding corolla, 2–3 mm. long
    in flower; corolla-lobes 7–8 mm. long · · · · · · · · · · · ·      18. *A. elegantulus*
    Leaves blunt; basal swelling of corolla 3.5–6 mm. across;
    calyx truncate and split into several segments by
    expanding corolla-base or if unilaterally split (subsp.
    *montanus*) then corolla-lobes 10–11 mm. long · · · · · ·      20. *A. sansibarensis*
24. Corolla-lobes not much hardened inside, the upper
    expanded part twice as long as the ill-formed claw,
    villous or lanate outside; paler, hardened, smooth distal
    part of filaments ± as long as basal part; filaments
    extended down corolla-tube as broad ± red wavy bands      26. *A. molleri*
    Corolla-lobes hardened inside, upper part subequal to
    folded claw; filaments without or with only a short
    smooth tip; filament-lines narrow down inside of tube · · · · · · · · · · · · · · 25
25. Corolla 4–6 cm. long, the lobes 9–11 mm. long · · · · · · · · · · · · · · · · · · · · 26
    Corolla 3.5–4.2 (–4.5) cm. long, the lobes 7–9 mm. long · · · · · · · · · · · · · 31
26. Corolla glabrous to thinly puberulous · · · · · · · · · · · · · · · · · · · · · · · · · · · 27
    Corolla densely covered with short to long hairs · · · · · · · · · · · · · · · · · · · 29
27. Corolla generally glabrous or if with hairs (Ufipa area)
    then all small and similar; corolla-buds pointed · · · ·      23. *A. nyasicus*
    Corolla minutely puberulous on bud-head and with
    scattered short ± branched hairs on tube; corolla-buds
    blunt · · · · · · · · · · · · · · · · · · · · · · · · · · · · · · · · · · · · · · · · · · · · · · · · · 28

28. Calyx 0.7–1 mm. long; branchlets pubescent with short
    spreading hairs, the longer ones branched and red-
    brown; peduncle 3–5 mm. long ················     24. *A. entebbensis*
    Calyx 0.3–0.5 mm. long; branchlets puberulous with very
    short spreading hairs, soon glabrescent; peduncle 7–10
    mm. long ······································     25. *A. keilii*
29. Corolla-lobes markedly keeled, the lobes thus thicker
    than broad ····································     30. *A. combreticola*
    Corolla-lobes not or slightly keeled, broader than thick ···············     30
30. Corolla velvety with minute red-brown branched hairs,
    the individual hairs scarcely distinguishable with a ×10
    lens ·········································     27. *A. songeensis*
    Corolla densely and coarsely hairy, the hairs mostly
    branched, longer ones wispy ··················     28. *A. musozensis*
31. Corolla-lobes each with a prominent appendage just
    below tip ·····································     33. *A. atrocoronatus*
    Corolla-lobes not spurred ·······························     32
32. Bracts much exceeding receptacle, 4–12 mm. long,
    ultimately ± slightly foliaceous; indumentum of
    branches, flowers and underside of young leaves very
    short, brown and much-branched, almost scurfy ····     29. *A. uhehensis*
    Bracts shorter to scarcely longer than receptacle, ovate-
    triangular; indumentum not scurfy ·····················     33
33. Branchlets glabrescent within a few internodes, soon
    densely lenticellate ···························     34
    Branchlets persistently pubescent ± the length of the
    flowering branchlets, not obviously lenticellate ·············     35
34. Corolla-lobes keeled; apex of buds subtruncate to
    rounded, ribbed, puberulous; corolla yellow, banded
    orange around vents, becoming dark red at tip of
    lobes ········································     31. *A. scassellatii*
    Corolla-lobes rounded on back; apex of buds subacute,
    terete, minutely scurfy-puberulous only at tip; corolla
    deep orange to fiery red ······················     32. *A. igneus*
35. Apical swelling of corolla-buds fusiform-conical,
    sometimes slightly ribbed; leaves grey-green to
    markedly blue-green; corolla with generally
    conspicuous longer hairs on constriction above the
    basal swelling ································     34. *A. pungu*
    Apical swelling of corolla-buds ellipsoid or oblong-
    ellipsoid, pentagonal; leaves dull green; corolla-hairs
    less differentiated ···························     35. *A. schweinfurthii*

1. **A. unyorensis** (*Sprague*) *Polh. & Wiens* in Lebrun & Stork, Énum. Pl. Fl. Afr.
Trop. 2: 165 (1992) & Mistletoes Afr.: 146 (1998). Type: Uganda, Bunyoro District,
above Kibiro [Kibero], *Bagshawe* 911 (BM!, holo.)

Shrubs with spreading and pendent branches, glabrous. Leaves opposite to
alternate; petiole 5–20 mm. long; lamina thinly coriaceous, green, glossy, lanceolate
to linear-lanceolate or elliptic-lanceolate, 8–14 cm. long, 2.5–5 cm. wide, acuminate
at the apex, basally cuneate to rounded or truncate, with 8–12 pairs of lateral nerves,
the lower to middle ones more strongly ascending. Heads 1–several per axil,
10–16(–24)-flowered; peduncle (plus rhachis) 4–8 mm. long; pedicels 3–5(–7) mm.
long; bract saucer-shaped with a small triangular limb, 1.5–2 mm. long. Receptacle
obconic, 1.5–2 mm. long; calyx 0.7–1 mm. long. Corolla 4.5–6 cm. long, pinkish red
to mauve with yellow to orange band around vents, the reddish tip only slightly

swollen over the anthers, apiculate, not swollen at the base; tube papillate on upper part inside, lobes erect, ± coherent, 10–12 mm. long, the expanded part linear-lanceolate, 5–7 mm. long, 1 mm. wide. Stamens inserted near top of tube, inflexed to involute, reddish, tip of filaments slightly narrowed and hardened; no tooth; anthers 3–3.5 mm. long. Style slightly broadened opposite the filaments, with a slight neck 3.5–5 mm. long; stigma ovoid, 0.8 mm. across. Berry white, marked red, oblong-obovoid, 10 mm. long, 6 mm. in diameter; seed orange.

UGANDA. Bunyoro District: Budongo Forest, Nov. 1939, *Eggeling* 3834!; Mengo District: Entebbe, 1918, *Fyffe* in *F.D.* 27! & 28! & Kijude, Nov. 1915, *Dummer* 2644!
TANZANIA. Bukoba District: Rusinga Forest Reserve, 24 Nov. 1994, *Congdon* 388!
DISTR. U 2, 4; T 1; Cameroon, Zaire and Burundi
HAB. Rain-forest, on various hosts; 900–1200 m.

SYN. *Loranthus unyorensis* Sprague in F.T.A. 6(1): 330 (1910) & in K.B. 1911: 144 (1911)
    *L. flamignii* De Wild. in B.J.B.B. 4: 413 (1914) & 5: 201 (1916); Balle in F.C.B. 1: 327 (1948).
        Type: Zaire, Eala, *Flamigni* 40A (BR!, holo.)
    *Tapinanthus flamignii* (De Wild.) Danser in Verh. K. Akad. Wet., sect. 2, 29(6): 112 (1933)
    *T. unyorensis* (Sprague) Danser in Verh. K. Akad. Wet., sect. 2, 29(6): 121 (1933)
    *Agelanthus flamignii* (De Wild.) Balle in B.J.B.B. 37: 457 (1967)
    [*Tapinanthus platyphyllus* sensu Balle in Fl. Cameroun 23: 50, t. 11/1–4 (1982), *non* (A. Rich.) Danser]

2. **A. platyphyllus** (*A. Rich.*) *Balle* in Stuttg. Beitr. Natur. 221: 2 (1970); Polh. & Wiens in U.K.W.F., ed. 2: 156 (1994) & Mistletoes Afr.: 147, photo. 49 (1998). Type: Ethiopia, Tigray, Djeladjeranne, *Schimper* III. 1584 (P!, holo., B!, BM!, BR!, FT!, G!, K!, L!, P!, iso.)

Shrub with spreading and pendent stems to 1 m. or so, glabrous. Leaves opposite to alternate; petiole 8–20 mm. long; lamina thinly coriaceous, glaucous, mostly lanceolate to oblong-elliptic, ovate or ovate-elliptic, 4.5–14 cm. long, 2.5–6 cm. wide, subacute to obtuse at the apex, basally cuneate to rounded, the lower ones on each branch smaller and more obovate, with 6–12 pairs of lateral nerves, the middle ones more ascending. Heads solitary in the axils, 10–16-flowered; peduncle 1–1.8 cm. long; pedicels 5–10 mm. long; bract saucer-shaped, with a small pointed or truncate triangular limb, 1.5–2 mm. long. Receptacle obconic, 1.5–2 mm. long; calyx 0.5–1 mm. long. Corolla 3.5–5 cm. long, brilliant red to orange-red with paler interior, only slightly swollen over the anthers, apiculate, not swollen at the base, verrucose inside on upper part of the tube; lobes longer than tube, erect, coherent, 2–2.7 cm. long, the expanded part linear, tapered above, (12–)15–16 mm. long, 1–1.2 mm. wide. Stamens inserted on lobes near top of claw, involute, tip of filaments slightly narrowed and hardened; no tooth; anthers 7–8.5 mm. long. Style virtually isodiametric; stigma capitate, 0.8 mm. across. Berry obovoid, flat-topped, 10 mm. long, 9 mm. in diameter. Fig. 9.

UGANDA. Acholi District: Chua, Lachung, Mar. 1935, *Eggeling* 1709!; Karamoja District: Moruita, Jan. 1959, *Tweedie* 1770!; Teso District: 0.5–1 km. E. of Mt. Alekilek, 10 May 1970, *Lye* 5447!
KENYA. Nandi District: 8 km. S. of Nandi Hills, 8 Mar. 1974, *Davidse* 7125!; Kisumu-Londiani District: Songhor, 19 Apr. 1965, *Archer* 497!; Kericho District: 8 km. N. of Chemelil on Nandi Hills road, 14 Mar. 1985, *Kirkup* 38!
DISTR. U 1, 3; K 3, 5; Sudan and Ethiopia
HAB. Deciduous woodland and wooded grassland, almost always on *Combretum* or *Terminalia*, 1200–1600 m.

SYN. *Loranthus platyphyllus* A. Rich., Tent. Fl. Abyss. 1: 341 (1848); Sprague in F.T.A. 6(1): 330 (1910); V.E. 3(1): 93, fig. 55 (1915); Engl. & K. Krause in E. & P. Pf., ed. 2, 16B: 154, fig. 72 (1935); F.P.S. 2: 293 (1952); Cufod., E.P.A.: 30 (1953) & in Senck. Biol. 39: 107 (1958); U.K.W.F.: 330 (1974)

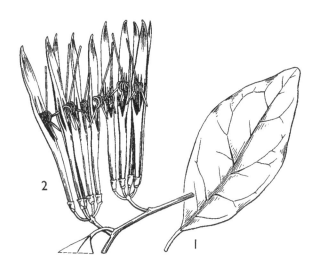

FIG. 9. *AGELANTHUS PLATYPHYLLUS* — **1**, leaf, × 1; **2**, inflorescence, × 1. From *Mooney* 6822. Drawn by Eleanor Catherine. Reproduced from Flora of Ethiopia.

*Schimperina platyphylla* (A. Rich.) Tiegh. in Bull. Soc. Bot. Fr. 42: 257 (1895); M.G. Gilbert
in Fl. Ethiopia 3: 372, fig. 114.13 (1990)
*Tapinanthus platyphyllus* (A. Rich.) Danser in Verh. K. Akad. Wet., sect. 2, 29(6): 117 (1933)

NOTE. The exceptional length of the corolla-lobes, with the filaments attached on the lobes, is
unmatched elsewhere in *Agelanthus* and the reason that M.G. Gilbert resurrected the generic
status of *Schimperina*. If the species is correctly allied with *A. unyorensis*, and it has even been
combined with it by Balle (1982), then it is correctly included within *Agelanthus*.

3. **A. microphyllus** *Polh. & Wiens*, Mistletoes Afr.: 150 (1998). Type: Tanzania,
Kondoa District, Kolo, *Polhill & Paulo* 1142 (K!, holo., BR!, EA!, P!, SRGH!, iso.)

Glabrous shrub with spreading or tangled branches to 1 m. or so, with tufts of
leaves and flowers pointing upwards. Leaves subopposite to alternate on long shoots,
mostly in tufts on very short shoots, shortly petiolate, linear to oblanceolate, 5–30
mm. long, 0.7–4 mm. wide, pointed to rounded at the apex, attenuate at the base,
obscurely nerved. Umbels terminal from leaf-clusters, 2–4-flowered; peduncle 0–7
mm. long; pedicels 3–10 mm. long; bract with a small lanceolate-triangular limb 1–3
mm. long. Receptacle 1–1.5 mm. long; calyx 0.3–0.8 mm. long. Corolla 4–5 cm. long;
tube grey to slightly purplish grey, dull purplish or brownish inside, lobes marked
yellow or yellow and red on claws; apical swelling of buds narrowly elliptic, pointed
to slightly apiculate, 4–5 mm. long, 1.5–2 mm. in diameter; base of tube not swollen;
lobes erect, 9–10 mm. long, expanded around filament-attachment below the narrow
claw, upper half to two-thirds narrowly spathulate to linear-oblanceolate, attenuate to
tip. Stamens yellow, inflexed, tapered, narrowed near tip, without a tooth; anthers
1.8–2.5 mm. long. Style mostly green, yellow near the apex, slender, flattened above,
slightly narrowed to 4 mm. long neck below small green stigma. Berry obovoid, 10
mm. long, 5–6 mm. in diameter.

KENYA. Northern Frontier Province: ± 5 km. S. of Tuum on W. side of Mt. Nyiru, 28 Oct. 1978,
*M.G. Gilbert, Gachathi & Gatheri* 5188!; Teita District: Tsavo East National Park, Voi R. near
main entrance, 22 Sept. 1962, *R. Williams & Sheldrick* TNP/E97!; Kwale District: Vigurangani
Scheme, 20 Nov. 1958, *Moomaw* 1033A!

TANZANIA. Kondoa District: Kolo, 1 Feb. 1928, *B.D. Burtt* 1301! & 12 Jan. 1962, *Polhill & Paulo* 1112!
DISTR. **K** 1, 7; **T** 5; not known elsewhere
HAB. Deciduous bushland, on *Acacia*; 30–1500 m.
NOTE. *A. microphyllus* is related to *A. natalitius* (Meisn.) Polh. & Wiens and *A. gracilis* (Tölken & Wiens) Polh. & Wiens from southern Africa. The contraction of the short shoots has proceeded much further in the East African species.

4. **A. subulatus** (*Engl.*) *Polh. & Wiens* in Lebrun & Stork, Énum. Pl. Fl. Afr. Trop. 2: 165 (1992) & Mistletoes Afr.: 150, photo. 3, 52, fig. 12 (1998). Types: Tanzania, Lushoto District, Silai, *Holst* 2299 (B!, syn., COI!, K!, isosyn.) & Lutindi, *Holst* 3304 (B!, syn., K!, isosyn.) & Kwa Mshusa, *Holst* 4117 (B!, syn.)

Stems spreading to 1–2 m. long, often with rather short lateral branches; branchlets flattened, becoming terete or 4-angled to slightly winged; plant glabrous except sometimes for short hairs in axils, on umbels and bracts. Leaves opposite or subopposite; petiole (0–)2–6(–10) mm. long; lamina thinly coriaceous, dull to dark green or sometimes slightly blue-green, mostly lanceolate to ovate-elliptic or elliptic, 4–12 cm. long, 2–5 cm. wide, lowest ones on flowering branchlets smaller and more obovate, narrowed or slightly acuminate to the obtuse or subacute apex, cuneate to rounded or rarely cordate at the base, with 6–12 pairs of lateral nerves, 2–several lower ones more strongly ascending. Umbels solitary in the axils or terminal on short shoots, 8–16-flowered; peduncle 0.8–2.5(–4) cm. long; pedicels 1–3 mm. long; bracts shallowly cupular with an enlarged limb; limb lanceolate to oblanceolate or foliaceous, flat or boat-shaped, sometimes slightly to conspicuously spurred from the back, 6–20 mm. long, 3–5 mm. wide. Receptacle 1–1.5 mm. long; calyx 0.6–1 mm. long. Corolla 3.5–5(–5.5) cm. long, white at the base, then pink or orange, banded green-yellow just below vents, yellow or yellow and red over vents, with green tip, occasionally almost wholly white; apical swelling obovoid-ellipsoid, slightly apiculate, angular to slightly ribbed, 3.5–5 mm. long, 2–2.5 mm. in diameter; basal swelling slight; lobes erect, 8–10(–12) mm. long, the upper expanded part oblong-elliptic, 3.5–4(–5) mm. long, 1–1.2(–1.5) mm. wide. Stamen-filaments red below, yellow or green above; tooth 0.5–0.8 mm. long; anthers 1.5–2(–3) mm. long. Style green above, slender, isodiametric; stigma capitate, 0.6 mm. across. Berry red, oblong-ellipsoid, 7–8 mm. long, 4.5–5 mm. in diameter; seed bright yellow. Fig. 10/1–6.

KENYA. N. Nyeri District: 7 km. on Nyeri–Kiganjo road, Zawadi Estate, 2 June 1974, *Faden & Evans* 74/706!; Kilifi District: S. end of Sokoke Forest 6 km. W. of Mombasa–Malindi road, 23 Sept. 1982, *Polhill* 4823!; Lamu District: 3 km. N. of Baragoni on road to Ijara and Kiunga, 4 Mar. 1980, *M.G. Gilbert & Kuchar* 5853!
TANZANIA. Lushoto, 20 Nov. 1962, *Semsei* 3524!; Morogoro District: Uluguru Mts., Kitundu, 1 Dec. 1934, *E.M. Bruce* 213!; Lindi District: Rondo Forest Reserve, 6 km. N. of Forest Station, 11 Feb. 1991, *Bidgood, Abdallah & Vollesen* 1482!
DISTR. **K** 4, 7; **T** 3, 6–8; Zaire (Shaba), S. Somalia, Zambia, Malawi, Mozambique and Zimbabwe
HAB. Forest to woodland, bushland and wooded grassland, generally on legumes, but sometimes on *Combretum* and genera of other families, sometimes parasitic on other mistletoes, may also attack cultivated trees; 10–2300 m.

FIG. 10. *AGELANTHUS SUBULATUS* — 1, flowering node, × 1; **2**, flower, × 2; **3**, stamen, × 4; **4**, style-tip, × 4; **5**, fruits, × 3; **6**, seed, × 3. *AGELANTHUS ZIZYPHIFOLIUS* subsp. *VITTATUS* — **7**, flowering branch, × 1; **8**, flower bud, × 2; **9**, flower, × 2; **10**, stamen, × 4; **11**, style-tip, × 4; **12**, fruit, × 4. *AGELANTHUS ELEGANTULUS* — **13**, flowering node, × 1; **14**, flower, × 2; **15**, anther, × 4; **16**, style-tip, × 4; **17**, fruit, × 2. *AGELANTHUS BIPARTITUS* — **18**, base of flower, showing circumscissile calyx, × 2. *AGELANTHUS SCASSELLATII* — **19**, flowering branch, × 1; **20**, tip of corolla-lobe, × 4; **21**, section of base of flower, × 4; **22**, stamen, × 4; **23**, style-tip, × 4. 1–6, from *Milne-Redhead & Taylor* 8507; 7–11, from *Congdon* 141; 12, from *Polhill & Pope* 4756; 13, 14, from *Drummond & Hemsley* 2806; 15, 16, from *Polhill et al.* 4979; 17, from *Gillett* 20821; 18, from *Milne-Redhead & Taylor* 11025; 19–23, from *Bally & Smith* 14938. Drawn by Christine Grey-Wilson.

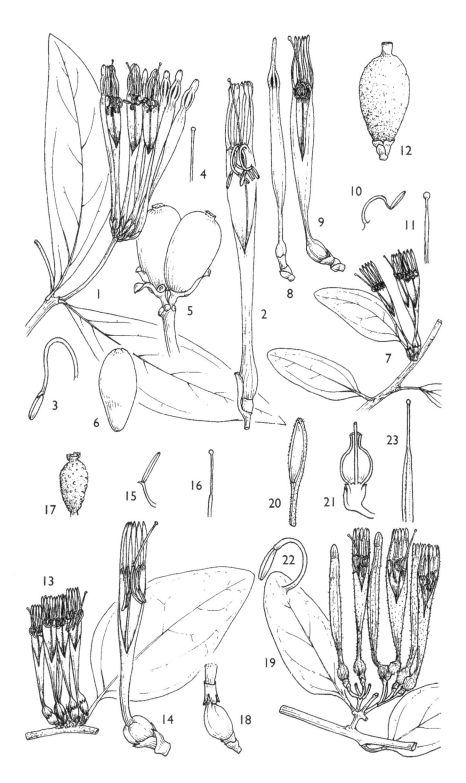

SYN. *Loranthus subulatus* Engl. in E.J. 20: 112 (1894) & P.O.A. C: 166, t. 17B–D (1895); Sprague
in F.T.A. 6(1): 330 (1910); Engl. in V.E. 3(1): 100, fig. 65 (1915); F.D.O.-A. 2: 171 (1932);
Engl. & K. Krause in E. & P. Pf., ed. 2, 16B: 165, fig. 83 (1935); T.T.C.L.: 292 (1949)

*L. usambarensis* Engl. in E.J. 20: 111 (1894) & P.O.A. C: 166 (1895); Sprague in F.T.A. 6(1):
340 (1910); F.D.O.-A. 2: 171 (1932); T.T.C.L.: 292 (1949). Types: Tanzania, Tanga
District, Moa [Muoa], *Holst* 3130 (B!, syn., K!, fragment) & Uzaramo District, Vikindu,
*Stuhlmann* 6132 (B!, syn.)

*L. latibracteatus* Engl. in E.J. 30: 303 (1901); Sprague in F.T.A. 6(1): 340 (1910); F.D.O.-A.
2: 171 (1932); T.T.C.L.: 291 (1949). Type: Tanzania, Rungwe District, Mt. Ntuli, *Goetze*
1310 (B!, holo., K!, fragment)

*L. thomasii* Engl. in E.J. 40: 531 (1908); F.D.O.-A. 2: 171 (1932). Type: Kenya, Kilifi District,
Takaungu, *F. Thomas II.* 63 (B!, holo., BR!, iso., K!, fragment)

*L. chunguensis* R.E. Fr., Wiss. Ergebn. Schwed. Rhod.-Kongo-Exped. 1: 24, fig. 4 (1914);
Balle in F.C.B. 1: 335 (1948), as '*chungwensis*'. Type: Zambia, N. Province, Chungu, *R.E.
Fries* 1194 (UPS, holo., K!, iso.)

*Tapinanthus chunguensis* (R.E. Fr.) Danser in Verh. K. Akad. Wet., sect. 2, 29(6): 110 (1933)

*T. latibracteatus* (Engl.) Danser in Verh. K. Akad. Wet., sect. 2, 29(6): 115 (1933)

*T. subulatus* (Engl.) Danser in Verh. K. Akad. Wet., sect. 2, 29(6): 120 (1933)

*T. usambarensis* (Engl.) Danser in Verh. K. Akad. Wet., sect. 2, 29(6): 121 (1933)

NOTE. Very attractive and readily recognised species, but also variable in some features that
suggest incipient divergence. The subdivision into three species by Sprague (1910) has some
basis that is still apparent. The northern plants as far south as Morogoro have distinctly angled
to slightly winged branchlets. Among these the inland plants from the Pare and Usambara Mts.
south to the Uluguru foothills have the bracts spurred to varying degrees (*L. subulatus*), a feature
lacking in the coastal plants (*L. usambarensis*, which, despite the epithet, was described from the
coast). With little exception the common element of the *Brachystegia* woodlands and
submontane forests further south have branchlets that are slightly flattened at first, but not
notably angular, and in addition more branchlets tend to be shortened, some not infrequently
bearing terminal inflorescences (the diagnostic feature of *L. chunguensis*). The bracts vary freely
in size and are occasionally spurred, and in populations from around the Livingstone Mts., NE.
of Lake Malawi, relatively small and hairy. The branchlets of this population are somewhat
angular and come close to the type locality of *L. latibracteatus*. No branchlets remain on the
fragment of the type seen, but Sprague describes them as subtetragonal; the bracts are, however,
characteristic of the common southern form. Other striking variations have been collected, such
as *Schlieben* 6172, from SE. Tanzania, Muera Plateau, and *Correia* 98, from N. Mozambique, Santa
Antonio Mission near Mueda in Caba Delgado, with sessile shortly cordate leaves.

5. **A. longipes** (*Baker & Sprague*) Polh. & *Wiens* in Lebrun & Stork, Énum. Pl. Fl.
Afr. Trop. 2: 164 (1992) & Mistletoes Afr.: 151 (1998). Type: Tanzania, Uzaramo
District, Dar es Salaam, *Kirk* (K!, holo., B!, iso.)

Stems spreading and pendent to 1 m. long, often with rather short lateral
branches; branchlets slightly compressed at first, soon terete; plant glabrous. Leaves
opposite or subopposite; petiole 3–7 mm. long; lamina thinly coriaceous, dark green,
mostly ovate-elliptic to elliptic, 3–11 cm. long, 2–6.5 cm. wide, lowest on side
branches smaller and more obovate, narrowed or acuminate to acute or shortly
rounded apex, cuneate to rounded at the base, with 6–12 pairs of lateral nerves, the
lower ones stronger and more ascending. Umbels axillary or terminal on short
branches, 4–8(–14)-flowered; peduncle 1.5–4 cm. long; pedicels 2–4 mm. long; bract
saucer-shaped with a small triangular-ovate limb 1.5–2.5 mm. long. Receptacle 1–1.5
mm. long; calyx 0.2–0.5 mm. long. Corolla 3.5–4.5 cm. long, tube pink or pinkish
mauve (sometimes white below), often darker lined, then green with head yellow or
green in bud, ± darker marked around vents; lobes green inside; apex of bud ovoid-
ellipsoid to ellipsoid, angled, slightly ribbed, pointed or slightly acuminate at the
apex, 3–4 mm. long, 2–2.5 mm. in diameter; basal swelling slight; lobes erect, 10–11
mm. long, the upper expanded part narrowly elliptic, 4 mm. long, 1.2 mm. wide.
Stamen-filaments dark red or purple; tooth 0.5–0.7 mm. long; anthers 2–2.5 mm.
long. Style green, slender, only slightly narrowed above; stigma capitate, 0.7–0.8 mm.
across. Berry red, not seen.

KENYA. Kwale District: Kaya Funga, 19 June 1994, *Luke* 4023! & Shimba Hills, Kwale Forest Reserve, 2 May 1968, *Magogo & Glover* 968!; Kilifi District: Arabuko-Sokoke Forest, N. of Sokoke Forest Station, 8 June 1973, *Musyoki & Hansen* 999!
TANZANIA. Tanga District: Pongwe, 23 May 1966, *Faulkner* 3788!; Uzaramo District: Pugu Hills, 18 June 1983, *Polhill & Lovett* 4878!; Lindi District: Rondo Plateau, St. Cyprians College, 17 Feb. 1991, *Bidgood, Abdallah & Vollesen* 1623!
DISTR. **K** 7; **T** 3, 6, 8; Mozambique
HAB. Coastal forest and associated bushland; 30–650 m.

SYN. *Loranthus longipes* Baker & Sprague in F.T.A. 6(1): 341 (1910) & in K.B. 1911: 146 (1911); F.D.O.-A. 2: 172 (1932); T.T.C.L.: 291 (1949)
  *Tapinanthus longipes* (Baker & Sprague) Danser in Verh. K. Akad. Wet., sect. 2, 29(6): 115 (1933)


6. **A. rondensis** (*Engl.*) *Polh. & Wiens* in Lebrun & Stork, Énum. Pl. Fl. Afr. Trop. 2: 165 (1992) & Mistletoes Afr.: 153 (1998). Type: Tanzania, Lindi District, eastern slopes of Rondo Plateau, *Busse* 2554 (B!, holo., BM!, BR!, EA!, iso., K!, fragment)

Plant glabrous; branchlets angular, densely lenticellate. Leaves opposite or subopposite; petiole 5–10 mm. long; lamina coriaceous, oblong-lanceolate or ovate-oblong, 3–7 cm. long, 1.5–3.5 cm. wide, acute to obtuse at the apex, cuneate to rounded at the base, penninerved with 4–6 lateral nerves on each side. Umbels 2–4-flowered; peduncle 1–1.5 mm. long; pedicels 0.5 mm. long; bract cupular, with a short truncate limb, 1–2 mm. long. Receptacle 0.7 mm. long; calyx saucer-shaped, 0.5–0.7 mm. long, ciliolate. Corolla 2.5 cm. long or ?rather more, with yellow basal swelling, otherwise purplish red; apical swelling of bud slightly oblong-fusiform, slightly apiculate, 4.5 mm. long, 1.5 mm. in diameter; basal swelling 2–2.5 mm. long, 1.5 mm. in diameter, with tube narrowly constricted for 2 mm. above; lobes erect, 9–10 mm. long, upper half narrowly linear-oblanceolate, only slightly hardened inside. Filaments inserted a little above the base of the lobes, slender, decurved, slightly corrugated; anthers 3 mm. long; connective produced apically with a transversely elliptic appendage 0.4 mm. long and as wide as the anther. Style slender; stigma globose, 0.5 mm. across. Berry not seen.

TANZANIA. Lindi District: Rondo Plateau, 16 May 1903, *Busse* 2554!
DISTR. **T** 8; known only from the type gathering
HAB. Not recorded

SYN. *Loranthus rondensis* Engl. in E.J. 40: 524 (1908); Sprague in F.T.A. 6(1): 327 (1910); F.D.O.-A. 2: 169 (1932) & T.T.C.L.: 289 (1949), excl. distr. E. Usambara
  *Tapinanthus rondensis* (Engl.) Danser in Verh. K. Akad. Wet., sect. 2, 29(6): 118 (1933)

NOTE. A diminutive but distinctive species most closely related to several species in south-central and southern Africa. These are the species of *Agelanthus* most similar to *Oncocalyx*.


7. **A. kayseri** (*Engl.*) *Polh. & Wiens* in Lebrun & Stork, Énum. Pl. Fl. Afr. Trop. 2: 164 (1992); Vollesen in Opera Bot. 59: 63 (1980), *nom. invalid.*; Polh. & Wiens in U.K.W.F., ed. 2: 156 (1994) & Mistletoes Afr.: 154, fig. 13B (1998). Types: Tanzania, Pangani, *Stuhlmann* I. 116 (B!, syn., K!, fragment) & Uzaramo District, Dar es Salaam, *Hildebrandt* 1226 (B†, syn., BM!, K!, P!, isosyn.)

Shrub to 1 m. or so, glabrous; branchlets slightly compressed, then slightly 4-angular. Leaves opposite to alternate; petiole 0–2 mm. long; lamina dull green to yellow-green, fleshy, ovate to ovate-elliptic or obovate, 3–8 cm. long, 2–6 cm. wide, obtuse or rounded at the apex, cuneate to attenuate to slightly cordate at the base, 3–5(–7)-nerved from a little above the base. Umbels 1–several, mostly at older nodes, 2–4-flowered; peduncle 1–2 mm. long; pedicels 1 mm. long; bract cupular with a small ovate-triangular limb, 1–2.5 mm. long. Receptacle 1.5 mm. long; calyx tubular, 2–2.5 mm. long. Corolla 1.8–3.3 cm. long, pink, lobes green with upper two-thirds red, or sometimes all cream or greenish cream; apical swelling of bud slightly clavate, 3–5 mm. long, 1.7–2 mm. in diameter, apiculate, slightly 5-angled to sutures above and to keels below; basal swelling 0; lobes erect, 6–11 mm. long, keeled, upper two-thirds linear-oblanceolate, apiculate, slightly hardened inside. Filaments yellow-green or cream, inflexed, slightly corrugated, last 1.5–2 mm. thinner and smooth; anther 2–3.5 mm. long. Style slender, minutely papillate opposite the upper part of corolla-tube, green above; stigma depressed-globose, 0.8 mm. across, red. Berry pink to red, obovoid-ellipsoid, 7–8 mm. long, 4–5 mm. in diameter, with persistent calyx.

Kenya. Kitui District: Endau Boma, 6 Aug. 1961, *Archer* 264!; Kilifi District: Ras Ngomeni, 15 Nov. 1961, *Polhill & Paulo* 767!; Tana R. District: 25 km. N. of Garsen on Garissa road, 15 Jan. 1972, *Wiens* 4521!

Tanzania. Tanga District: Kigombe beach, 12 July 1953, *Drummond & Hemsley* 3259!; Uzaramo District: Kunduchi ruins, 13 Nov. 1969, *B.J. Harris* 3590!; Kilwa District: ± 3 km. N. of Kingupira, 19 Apr. 1975, *Vollesen* in *M.R.C.* 2254!; Zanzibar I., Chwaka, 25 Oct. 1932, *Vaughan* 2002!

Distr. **K** 4, 7; **T** 3, 6, 8; **Z**; S. Somalia

Hab. Coastal bushland and mangrove stands, extending inland along rivers, commonly on *Dobera, Salvadora* or mangroves; 0–100 m. (or possibly more, see note)

Syn. *Loranthus kayseri* Engl. in E.J. 20: 89 (1894) & P.O.A. C: 165, t. 13A–D (1895); Sprague in F.T.A. 6(1): 318 (1910); F.D.O.-A. 2: 168 (1932); T.T.C.L.: 289 (1949)

   [*L. crassissimus* sensu Engl. in E.J. 20: 122 (1894), pro parte quoad specim. ex Pangani, *non* Engl. sensu stricto]

   *L. rhodanthus* Chiov., Fl. Somala 2: 383, fig. 218 (1932). Type: Somalia, Obe, *Senni* 349 (FT, holo.)

   *Tapinanthus kayseri* (Engl.) Danser in Verh. K. Akad. Wet., sect. 2, 29(6): 114 (1933)

   *Loranthus sp. D* sensu U.K.W.F.: 332 (1974)

Note. A distinctive species with leaves often somewhat resembling the Salvadoraceae on which it is most commonly found. The species does extend inland along the river systems but one specimen, *Wingfield* 2991, said to be from the Lukwangule Plateau of the Uluguru Mts., at 2000 m., may have been mislabelled. Peter (1932) also records the species from the foot of the W. Usambara Mts. at Mombo, and more improbably from Mbulu (perhaps based on specimens of *A. zizyphifolius* subsp. *vittatus*).

8. **A. zizyphifolius** (*Engl.*) Polh. & *Wiens* in Lebrun & Stork, Énum. Pl. Fl. Afr. Trop. 2: 166 (1992) & in U.K.W.F., ed. 2: 156 (1994) & Mistletoes Afr.: 154 (1998). Type: Tanzania, Tabora District, Rubugua, *Stuhlmann* 493 (B!, holo., K!, fragment)

Shrub 40–100 cm. or so; branchlets slightly compressed to 4-angular, then terete, glabrous to red-brown tomentose. Leaves alternate to subopposite, rather densely set; petiole 1–2(–5) mm. long; lamina leathery to rather thick and brittle, dull yellowish green, oblong-oblanceolate to oblong-obovate or elliptic, 2.5–10(–15) cm. long, 1–4(–4.5) cm. wide, obtuse to shortly rounded at the apex, attenuate at the base, glabrous or some hairs towards margins and base, 3(–5)-nerved from a little above the base. Flowers clustered in axils or at old nodes; pedicels 0–3 mm. long; bract cupular with a small triangular limb, 3–4 mm. long, ciliolate and sometime puberulous. Receptacle 1.5 mm. long; calyx tubular, 3–5 mm. long, shortly to distinctly toothed, ciliolate. Corolla 3–4.5 cm. long, pink below, sometimes with a green or red band at top of tube, whitish over vents, with red tips to lobes; apical swelling of bud slight, oblong to oblong-fusiform, 4–5 mm. long, 1.5–2 mm. in diameter; basal swelling 4–7 mm. long, 2–4 mm. in diameter, with tube narrowly constricted for 3–4 mm. above; lobes erect, 8–9 mm. long, with the upper three-eighths to two-thirds linear-oblanceolate, pointed, slightly keeled and hardened inside. Filaments red to purple, inflexed to involute, slightly tapered, corrugated, last 0.5–0.8 mm. paler, smooth, hardened; anthers 2–2.5 mm. long. Style green or whitish, slender to slightly winged opposite the filaments (then with a slight neck below stigma); stigma ovoid-conic, 0.7–0.8 mm. across. Berry red, ellipsoid-obovoid, 8 mm. long, 7 mm. in diameter, with persistent calyx; seed pink or yellow.

subsp. **zizyphifolius**; Polh. & Wiens, Mistletoes Afr.: 155, photo. 55 (1998)

Branchlets red-brown pubescent to tomentose; leaves generally pubescent at base and a little above, near midrib and at margins.

KENYA. Nakuru District: Njoro, Egerton Farm, 6 Feb. 1985, *Kirkup* 29!; Machakos District: Kilima Kiu, 19 Nov. 1967, *Gillett* 18358!; Masai District: Magadi–Nairobi road, 6 km. SW. of Kiserian, 11 Dec. 1971, *Wiens* 4470!
TANZANIA. Arusha District: Engare Nanyuki, 30 Nov. 1966, *Richards* 21644! & Engare Nanyuki R., 7 Dec. 1966, *Richards* 21666!; Kigoma District: Kasye Forest, 21 Mar. 1994, *Bidgood, Mbago & Vollesen* 2880!
DISTR. K 1, 3, 4, 6; T 2, 4; Burundi
HAB. Edges of upland dry evergreen forest and associated bushland along the Rift Valley and its flanks, including *Tarchonanthus* associations, extending into deciduous bushland, woodland and riverine communities nearby, also extending into *Combretum* woodland and bushland in western Tanzania, commonly on *Rhus*, but also on *Acacia, Grewia, Combretum, Parinari* and various other hosts; 900–2600 m.

SYN. *Loranthus zizyphifolius* Engl. in E.J. 20: 92 (1894) & P.O.A. C: 165 (1895); Sprague in F.T.A. 6(1): 321 (1910); F.D.O.-A. 2: 169 (1932); T.T.C.L.: 289 (1949); U.K.W.F.: 332 (1974); Blundell, Wild Fl. Kenya: 66, t. 43/280 (1982)
    *Tapinanthus zizyphifolius* (Engl.) Danser in Verh. K. Akad. Wet., sect. 2, 29(6): 122 (1933); Blundell, Wild Fl. E. Afr.: 131, t. 547 (1987)


subsp. **vittatus** (*Engl.*) Polh. *& Wiens* in Lebrun & Stork, Énum. Pl. Fl. Afr. Trop. 2: 166 (1992) & Mistletoes Afr.: 155, photo. 56, fig. 13D (1998). Type: Tanzania, Mbeya District, Nsundas [Suntas Dorf], *Goetze* 1431 (B!, holo., BR!, P!, iso., K!, fragment)

Branchlets and leaves glabrous. Fig. 10/7–12 (p. 47).

TANZANIA. Mbulu, Feb. 1927, *B.D. Burtt* 5597!; Iringa District: Mufindi, Kibao, 5 Jan. 1986, *D. & C. Wiens & Congdon* 6546!; Songea District: Luhila, 18 Sept. 1956, *Semsei* 2476!
DISTR. T 2, 4, 7, 8; Zaire, Burundi, Zambia, Malawi, Mozambique and Zimbabwe
HAB. Montane forest, riverine forest, wooded grassland and woodlands nearby, on various hosts; 850–2000 m.

SYN. *Loranthus vittatus* Engl. in E.J. 30: 301 (1901); Sprague in F.T.A. 6(1): 324 (1910); F.D.O.-A. 2: 169 (1932); Brenan, T.T.C.L.: 289 (1949), pro parte, & in Mem. N.Y. Bot. Gard. 9: 64 (1954)
    *L. pallideviridis* Engl. & K. Krause in E.J. 51: 463 (1914); F.D.O.-A. 2: 173 (1932); T.T.C.L.: 290 (1949). Type: Tanzania, Rungwe District, Ulambya [Bulambia], Songwe valley, *Stolz* 1613 (B!, holo., BM!, BR!, K!, MO!, P!, iso.)
    *Tapinanthus pallideviridis* (Engl. & K. Krause) Danser in Verh. K. Akad. Wet., sect. 2, 29(6): 117 (1933)
    *T. vittatus* (Engl.) Danser in Verh. K. Akad. Wet., sect. 2, 29(6): 122 (1933)

NOTE. The montane forest form from Mbulu to the Nyika Plateau differs from subsp. *zizyphifolius* essentially only in the lack of hairs, but from southern Tanzania southwards the subspecies descends along rivers into the plateaux, and is very commonly found on *Syzygium*. This form tends to have shortly pedicellate (rather than sessile) flowers and the style tends to be slightly winged opposite the filaments creating a slight neck below the stigma. The populations from western Tanzania and Burundi need further study. The type is only slightly hairy and both subspecies occur in the Ruyigi Province of Burundi.


9. **A. brunneus** (*Engl.*) Balle *& Hallé* in Adansonia, sér. 2, 1: 233 (1962), pro parte; Tiegh. in Bull. Soc. Bot. Fr. 42: 246, 738 (1895), *nom. invalid.*; Balle in F.W.T.A., ed. 2, 1: 660 (1958), *nom. invalid.*; Balle in Bol. Soc. Brot., sér. 2, 38: 54, t. 1/5, 4/2 (1964); Berhaut, Fl. Ill. Sénégal 6: 59, fig. on 58 (1979); Polh. & Wiens, Mistletoes Afr.: 156, photo. 57 (1998). Type: Angola, Pungo Andongo, Pedras de Guinga and Mutollo, *Welwitsch* 4850 (B!, holo., COI!, G!, K!, LISU!, P!, iso.)

Stems to 1 m. or so; plants glabrous except sometimes slightly farinose around axils, branchlets angled. Leaves alternate to opposite and ternate; petiole 0–2 mm. long, the lamina expanding on the attenuate part at the base so the leaves become more sessile with age; lamina coriaceous, light to dark green or yellow-green, elliptic to oblanceolate or obovate, 2.5–13 cm. long, 1.5–8 cm. wide, varying considerably in size even on same plant, obtuse or rounded at the apex, cuneate and decurrent to base, oldest occasionally rounded, 3(–5)-nerved from base. Flowers few–numerous, crowded in axils and at older nodes; pedicels 0–1.5 mm. long; bract cupular to tubular, umbonate, with a very small limb, 1.5–3 mm. long. Receptacle 1.5–2 mm. long; calyx tubular, (2–)2.5–3 mm. long, ciliolate. Corolla 3.2–4 cm. long, pink or occasionally greenish white, with several coloured bands near the base of the lobes, the tips red, elsewhere corolla-tube sometimes yellow to orange; tube narrowed over filaments in bud, only slightly expanded over the anthers; basal swelling usually conspicuous from inception or at least from fairly young bud-stage, obovoid, attenuate to base, not splitting the calyx, 5–6 mm. long, 2–3 mm. in diameter, tube narrowly constricted for several mm. above; tube ± papillate only along sutures inside; lobes erect, 8–9 mm. long, upper part linear-elliptic, 5–6 mm. long, 0.6–0.8 mm. wide. Filaments red, incoiled once, flattened, tapered, corrugated with the last 0.5 mm. hardened; anthers 1.5–2 mm. long. Style green to red, slightly angled, scarcely swollen opposite the filaments, tapered above; stigma capitate, 0.6–0.7 mm. across. Berry obovoid to urceolate, 7–8 mm. long, 4.5–5 mm. in diameter, crowned by the calyx, smooth.

UGANDA. W. Nile District: Okollo, Bula Forest, Mar. 1935, *Eggeling* 1946!; Kigezi District: Impenetrable Forest, 30 Sep. 1970, *Katende* 570!; Mengo District: Mulange, Oct. 1919, *Dummer* 4337!

KENYA. N. Kavirondo District: S. Elgon, Sosian R., Sept. 1951, *Tweedie* 925! & near Kakamega, Rondo, 27 Aug. 1975, *G.C. van Someren* 131!; Kericho District: SW. Mau Forest, Sambret Catchment, 16 July 1962, *Kerfoot* 4056!

TANZANIA. Bukoba District: Minziro Forest, Mar. 1991, *Kielland* 11! & 12! & 14 & 15!

DISTR. **U** 1, 2, 4; **K** 5; **T** 1; westwards to Senegal and Angola

HAB. Evergreen forest, swamp and riverine forest; 1000–1800 m.

SYN. *Loranthus brunneus* Engl. in E.J. 20: 88 (1894); Hiern, Cat. Afr. Pl. Welw. 1: 931 (1900); Sprague in F.T.A. 6(1): 325 (1910); Balle in F.C.B. 1: 331 (1948); F.P.N.A. 1: 103 (1948)
      *Tapinanthus brunneus* (Engl.) Danser in Verh. K. Akad. Wet., sect. 2, 29(6): 109 (1933); Balle in Fl. Cameroun 23: 64 (1982), pro parte; Troupin, Fl. Pl. Lign. Rwanda: 378 (1982), pro parte

NOTE. *A. brunneus* and its relatives are fairly easily recognised in East Africa. The flowers are consistently pink and white, providing an easy separation from the montane species *A. krausei* and *A. pennatulus*, but in central Africa *A. brunneus* is much more variable, sometimes with yellow flowers as well. In bud, before the basal swelling of the buds develop, *A. brunneus* may be confused with *A. djurensis*, but in general has blunt leaves broadest above the middle rather than acuminate leaves broadest about the middle or below, the flowers are not in obvious umbels and never pedunculate, the bract is cupular to tubular, rather than ovate-triangular from a shallow base (but note *A. toroensis* does have a cupular bract), the calyx is also more tubular, the corolla is generally smaller and not so dark in colour.

10. **A. djurensis** (*Engl.*) *Polh. & Wiens* in Lebrun & Stork, Énum. Pl. Fl. Afr. Trop. 2: 163 (1992) & Mistletoes Afr.: 157 (1998). Type: Sudan, Jur, Great Wau R., *Schweinfurth* 1632 (B!, holo., K!, P!, iso.)

Glabrous shrub with stems 0.5–2 m. long; young branchlets subangular on upper part of internode, terete below, becoming terete. Leaves mostly opposite or subopposite; petiole apparently 2–8 mm. long, but the attenuate lamina expanding with age so that leaves become sessile; lamina coriaceous, dark green, paler beneath, ovate to elliptic, 5–14 cm. long, 2.5–7 cm. wide, ± acuminate at the apex, cuneate to rounded or cordate to attenuate base, 3-nerved from base. Umbels several in axils and at older nodes below, 3–6-flowered; peduncle 0–3 mm. long; pedicels 1–2 mm. long; bract ovate-triangular from a shallow saucer-shaped base, 1.5–2 mm. long. Receptacle 1–1.5 mm. long; calyx cupular, 1–2 mm. long, ciliate. Corolla 4–5 cm. long, deep crimson or wine-red (sometimes paler yellow-green when young), paler or greenish over vents, darker tipped; tube narrowed over the filaments, slightly expanded over anthers; basal swelling slight, not rupturing the calyx; tube papillate inside; lobes erect, 7–9 mm. long, upper expanded part linear-elliptic to linear-oblanceolate, 4–6 mm. long, 1 mm. wide. Filaments inflexed, flattened, slightly corrugated, the last 1 mm. hardened; anthers 2–2.5 mm. long. Style slightly expanded below stamens, forming only a slight neck, slightly fluted or winged; stigma capitate, 0.7 mm. across. Berry red or white, obovoid, 9 mm. long, 6 mm. in diameter, with a persistent calyx, sparsely covered in small pimples at least when dried.

UGANDA. Bunyoro, Butiaba flats near Kisansya, 16 Sep. 1969, *Lye* 4009!; Kigezi District: Ishasha, 26 Sept. 1968, *Lock* 68/214!; Mengo District: W. Wantuluntu, Jan. 1917, *Dummer* 3045!
DISTR. U 2, 4; Cameroon to Sudan and south to Angola
HAB. Evergreen forest; 950–1800 m.

SYN. *Loranthus djurensis* Engl. in E.J. 20: 90, t. 1C (1894); Sprague in F.T.A. 6(1): 320 (1910); Balle in F.C.B. 1: 328 (1948); F.P.S. 2: 293 (1952)
　　*L. trinervius* Engl. in E.J. 40: 527 (1908); Sprague in F.T.A. 6(1): 319 (1910). Types: Cameroon, Lolodorf, Mbanga Mt., *Staudt* 127 (B!, syn., G!, K!, isosyn.) & 368 (B†, syn., G!, K!, LISC!, P!, isosyn.)
　　*Tapinanthus djurensis* (Engl.) Danser in Verh. K. Akad. Wet., sect. 2, 29(6): 111 (1933)
　　*Tapinanthus trinervius* (Engl.) Danser in Verh. K. Akad. Wet., sect. 2, 29(6): 121 (1933)
　　*T. brunneus* (Engl.) Danser subsp. *djurensis* (Engl.) Balle in Fl. Cameroun 23: 66, t. 12/17–21 (1982) & in Troupin, Fl. Rwanda 1: 189 (1978), *nom. invalid.*; Troupin, Fl. Pl. Lign. Rwanda: 378 (1982)

NOTE. Generally in areas with more seasonal rainfall than those occupied by *A. brunneus*. Included as a subspecies of that species by Balle (1982), but usually easily distinguished by the different coloured flowers and lack of a marked swelling to the base of the corolla. Some specimens with young flowers may be confused as indicated in the notes under *A. brunneus*. *A. toroensis* is similar, but has a more cupular bract and longer calyx.

11. **A. toroensis** (*Sprague*) *Polh. & Wiens*, Mistletoes Afr.: 158, photo. 58 (1998). Type: Uganda, Toro District, Mpanga R. mouth, *Bagshawe* 1157 (BM!, holo., K!, iso.)

Glabrous shrub; young branchlets subangular at least on the upper parts of internodes, becoming terete. Leaves mostly opposite or ternate; petiole apparently 2–5 mm. long, but with age the attenuate part of lamina expanding so leaves become sessile; lamina coriaceous, lanceolate-elliptic to elliptic, (3–)5–16 cm. long, (1.5–)2–7 cm. wide, slightly acuminate to the pointed tip, attenuate at the base, more rounded to cordate with age, 3-nerved from base. Umbels sessile in axils and at older nodes, mostly 4–6-flowered; pedicels 1–2 mm. long; bract cupular, slightly erose and ciliolate but with virtually no limb; receptacle hidden by bract, 2–3 mm. long. Calyx deeply cupular, 3 mm. long, split by expanding corolla-base. Corolla 5–5.5 cm. long, wine-red, yellow to orange over vents, usually with a green band at top of tube and with tip dark red; tube narrowed over filaments, slightly expanded over anthers; basal swelling slight, elongate, 7–10 mm. long, 3 mm. in diameter, next 10–12 mm. slightly narrowed; tube slightly papillate along filament-lines inside; lobes erect, 8–9 mm. long, the upper expanded part linear elliptic to linear oblanceolate, 5–6 mm. long, 1 mm. wide. Filaments red, inflexed, flattened, the distal 0.6 mm. slightly hardened; anthers 2.5–3 mm. long. Style green, slightly fluted or winged in section, slightly narrowed opposite the anthers forming a slight neck; stigma capitate, 0.6–0.8 mm. across. Berry not seen.

UGANDA. Bunyoro District: Bubwe, June 1943, *Purseglove* 1594!; Kigezi District: Queen Elizabeth National Park, Ishasha–Mweya road near turning for Ishasha River Camp, 5 Jan. 1985, *Kirkup* 8!; Mengo District: Mabira Forest, Chagwe, Dec. 1922, *Maitland* 527!

KENYA. N. Kavirondo District: main road W. of Bungoma, Nov. 1971, *Tweedie* 4136!
DISTR. U 2, 1; K 5; easternmost Zaire and Burundi
HAB. Wooded grassland and forest edges, recorded on *Acacia, Crataeva, Dombeya* and *Rhus*;
    600–1400 m.

SYN. *Loranthus toroensis* Sprague in K.B. 1916: 180 (1916)
    *L. crataevae* Sprague in K.B. 1916: 179 (1916). Type: Uganda, Toro District, near Semliki
        R., *Bagshawe* 1303 (BM!, holo., K!, US!, iso.)
    *Tapinanthus crataevae* (Sprague) Danser in Verh. K. Akad. Wet., sect. 2, 29(6): 110 (1933)
    *T. toroensis* (Sprague) Danser in Verh. K. Akad. Wet., sect. 2, 29(6): 121 (1933)
    [*Loranthus demeusei* sensu F.P.N.A. 1: 103 (1948), *non* Engl.]

NOTE. Closely related to *A. djurensis*, but readily distinguished by the cupular bract and
    deeper calyx.

12. **A. krausei** (*Engl.*) *Polh. & Wiens* in Lebrun & Stork, Énum. Pl. Fl. Afr. Trop. 2:
164 (1992) & Mistletoes Afr.: 159, photo. 59, fig. 13C (1998). Type: Rwanda, Rugege
Forest, *Mildbraed* 882 (B!, holo., K!, fragment)

Shrub to 1 m. or so; branchlets compressed to 4-angular, then terete, papillose-
puberulous only around axils and sometimes on young internodes above, soon
glabrescent, densely lenticellate. Leaves alternate to opposite or ternate; petiole 1–6
mm. long; lamina coriaceous, green, narrowly oblong-elliptic to slightly lanceolate- or
oblanceolate-oblong, 4–12(–15) cm. long, 1.5–2.5(–7) cm. wide, acute to obtuse at the
apex, cuneate at the base, glabrous, obscurely penninerved, sometimes (varying on
same branch) with basal or lower pair of nerves strongly ascending, with 6–10 pairs of
lateral nerves. Flowers clustered in axils or at old nodes, subsessile; bract cupular with
small triangular limb, 3–4 mm. long, ciliolate. Receptacle 1–1.5 mm. long; calyx
tubular or funnel-shaped, 3–4 mm. long, shortly toothed, ciliolate. Corolla (3.5–)4–4.5
cm. long, yellow or pale orange, banded green and then red below vents, whitish over
vents, with red tips to lobes; apical swelling of bud slight, oblong to oblong-fusiform, 7
mm. long, 1.5 mm. in diameter; basal swelling 4–7 mm. long, 2.5–4 mm. in diameter,
with tube narrowly constricted for 3–4 mm. above; lobes erect, 8–9 mm. long, with
upper two-thirds linear-oblanceolate, pointed, slightly hardened inside. Filaments red,
slightly involute, slender, only slightly corrugated, with very short smooth pale
hardened tip; anthers 1.8–2 mm. long. Style red, greenish above, slender, slightly
tapered to stigma; stigma ovoid to capitate, 0.7–1 mm. long. Berry not seen.

UGANDA. Kigezi District: Luhiza, June 1951, *Purseglove* 3680! & Kigezi, Sept. 1936, *Eggeling* 3275!
TANZANIA. Bukoba District: Minziro, 30 Apr. 1994, *Congdon* 367!
DISTR. U 2; T 1; Rwanda (but see note under next species)
HAB. Forest, recorded on *Mimulopsis*; 1150–2400 m.

SYN. *Loranthus krausei* Engl. in E.J. 43: 311 (1909); Sprague in F.T.A. 6(1): 326 (1910); F.D.O.-A.
        2: 169 (1932), pro parte, excl. specim. ex Usambara
    *Tapinanthus krausei* (Engl.) Danser in Verh. K. Akad. Wet., sect. 2, 29(6): 114 (1933).
    [*Loranthus glomeratus* sensu Balle in F.C.B. 1: 330 (1948), pro parte, *non* Engl.]
    *Tapinanthus brunneus* (Engl.) Danser subsp. *krausei* (Engl.) Balle in Troupin, Fl. Rwanda 1:
        189, fig. 37/3 (1978); Troupin, Fl. Pl. Lign. Rwanda: 378, fig. 131/3 (1982), *nom.
        invalid., comb. non rite publ.*

NOTE. Peter in F.D.O.-A. 2: 169 (1932) cited with a query mark a specimen he had collected
    from the Kwamkuyu Falls near Amani, *Peter* 17098. This has not been seen, and probably no
    longer exists. No species of the *A. brunneus* complex is known from the E. Usambara Mts., so
    it is not easy to guess what it might have been.

13. **A. pennatulus** (*Sprague*) *Polh. & Wiens* in Lebrun & Stork, Énum. Pl. Fl. Afr.
Trop. 2: 164 (1992) & in U.K.W.F., ed. 2: 156, t. 55 (1994) & Mistletoes Afr.: 160
(1998). Lectotype, chosen by Polh. & Wiens (1998): Kenya, Kiambu District, Limuru,
*Scheffler* 308 (K!, lecto., BM!, P!, WAG!, isolecto.)

Shrub to 1 m. or so; branchlets slightly compressed to 4-angular, then terete, glabrous, ± densely lenticellate. Leaves alternate to opposite; petiole to 2(–4) mm. long; lamina thinly coriaceous, dark green, sometimes red-veined, elliptic-lanceolate to ovate-elliptic, 4–10 cm. long, 2.5–6 cm. wide, subacute to obtuse or shortly rounded at the apex, broadly cuneate to rounded at the base, glabrous, rather inconspicuously penninerved with lower pair of nerves stronger and ascending to upper half. Flowers clustered in axils and at older nodes below, subsessile; bract cupular with a small triangular limb, 2.5–3 mm. long, ciliolate. Receptacle 1–1.5 mm. long; calyx slightly funnel-shaped, 2–2.5 mm. long, subtruncate, ciliolate. Corolla (3.5–)4–5 cm. long, with yellow-orange tube banded green and orange-red at top, lobes paler pinkish white over vents, with red to purple tips; apical swelling of bud slight, oblong, 6 mm. long, 1.5 mm. in diameter; basal swelling 5–6 mm. long, 2.5–3 mm. in diameter, with tube narrowly constricted for 4–6 mm. above; lobes erect, 8–9 mm. long, with the upper two-thirds linear-oblanceolate, slightly keeled and slightly hardened inside. Filaments red, slender, involute, scarcely corrugated, but apical 0.5–0.7 mm. smooth, pale, hardened; anthers 2–2.5 mm. long. Style green, slender; stigma ovoid-conic, 0.8–1 mm. long. Berry not seen.

KENYA. Meru District: Nyambeni Hills, Thangatha R. at S. circular road to Maua, 10 Oct. 1960, *Polhill & Verdcourt* 278!; Embu Forest, 28 Sept. 1956, *Ossent* 173!; Kiambu District: Muguga N. Forest, 25 Sept. 1963, *Verdcourt & Howard* 3784!
TANZANIA. Moshi District: above Kilimanjaro Timbers, 14 Jan. 1993, *Grimshaw* 9471! & 9472!; Morogoro District: Uluguru Mts., Bunduki, Jan. 1927, *E.M. Bruce* 670! & Minguti Forest above Chenzema village, 12 Mar. 1986, *J. & J. Lovett, Pocs & Bidgood* 561!
DISTR. K 4; T 2, 6; not known elsewhere (but see note)
HAB. Montane moist and upland dry evergreen forest, recorded on *Olea, Strombosia* and various Rutaceae; 1650–2400 m.

SYN. *Loranthus pennatulus* Sprague in F.T.A. 6(1): 324 (1910) & in K.B. 1911: 143 (1911)
    *L. keniae* K. Krause in N.B.G.B. 8: 496 (1923). Type: Kenya, N. Nyeri District, Mt. Kenya, W. slopes, *R.E. & T.C.E. Fries* 926 (UPS, syn., K!, fragment) & 1415 (UPS, syn.)
    *Tapinanthus keniae* (K. Krause) Danser in Verh. K. Akad. Wet., sect. 2, 29(6): 114 (1933)
    *T. pennatulus* (Sprague) Danser in Verh. K. Akad. Wet., sect. 2, 29(6): 117 (1933); Blundell, Wild Fl. E. Afr.: 131, t. 471 (1987)
    [*Loranthus brunneus* sensu U.K.W.F.: 334, fig. on 331 (1974), *non* Engl.]
    *L. sp. C* sensu U.K.W.F.: 332 (1974), pro parte

NOTE. The specimens from the Uluguru Mts., cited above, are somewhat intermediate between this species and *A. krausei* and may need to be segregated in due course. This montane complex related to *A. brunneus* has a further component, known from fragmentary material only, on Mt. Mulanje in southern Malawi.


14. **A. dodoneifolius** (*DC.*) *Polh. & Wiens* in Lebrun & Stork, Énum. Pl. Fl. Afr. Trop. 2: 163 (1992) & Mistletoes Afr.: 162, photo. 60 (1998). Type: Senegal, *Perrottet* (G-DC!, holo., P!, iso.)

Well-branched shrub with spreading and pendent branches to 1.5 m. long, glabrous; branchlets slightly 3-angular or flattened, soon terete. Leaves opposite to alternate or ternate; petiole 3–10 mm. long; lamina coriaceous, usually glaucous but sometimes yellow-green to grey-green, linear-lanceolate to oblong-lanceolate, usually elongate, occasionally broader and even oblanceolate, often slightly curved, 5–15(–22) cm. long, 0.5–2(–4.5) cm. wide, acute to rounded at the apex, cuneate at the base, obscurely nerved, the middle nerves more ascending. Umbels (1–)2–3-flowered; peduncle 1–3(–5 including connective between pedicel-sockets) mm. long; pedicel-sockets oblique; bract cupular, 3–5 mm. long. Receptacle 3–4 mm. long; calyx tubular, 4–6 mm. long, ciliate and slightly erose, split by the expanding basal swelling of corolla. Corolla (3.5–)4.5–5(–5.5) cm. long, white at the base (sometimes with red filament-lines), flushed wine-red above, sometimes greenish white inside, tip sometimes becoming purplish; apical swelling slight, bluntly pointed; basal swelling ellipsoid, 5–8 mm. long, 3.5–4 mm. in diameter; lobes erect, 15–16 mm. long, upper expanded part linear-elliptic to linear-oblanceolate, 10–11 mm. long, 1.5 mm. in diameter, thickened. Filaments red, involute, stout, flattened; tooth 1 mm. long. Style dark red, 5-winged especially on the thickened part opposite vents in bud; neck 5–6 mm. long; stigma ellipsoid, 0.8 mm. across. Berry orange-red to red, obovoid-ellipsoid, 1–1.5 mm. long, 0.8–1 cm. in diameter, with ± persistent calyx.

UGANDA. W. Nile District: Madi, Metuli, Apr. 1940, *Eggeling* 3869!
DISTR   U 1; Senegal to Sudan
HAB. Wooded grassland, on many hosts, recorded on *Butyrospermum* in Uganda; probably ± 900 m.

SYN. *Loranthus dodoneifolius* DC., Prodr. 4: 303 (1830), as '*dodoneaefolius*', & Mém. Fam. Loranthacées: 29, t. 9 (1830), as '*dodoneaefolius*'; Engl. in E.J. 20: 108 (1894), as '*dodoneaefolius*'; Sprague in F.T.A. 6(1): 341 (1910), as '*dodoneaefolius*'; F.P.S. 2: 293 (1952)
    *Dentimetula dodoneifolia* (DC.) Tiegh. in Bull. Soc. Bot. Fr. 42: 265 (1895)
    *Tapinanthus dodoneifolius* (DC.) Danser in Verh. K. Akad. Wet., sect. 2, 29(6): 111 (1933); Balle in F.W.T.A., ed. 2, 1: 662 (1958) & in Bol. Soc. Brot., sér. 2, 38: 59, t. 2/1, 8/1 (1964); Berhaut, Fl. Ill. Sénégal 6: 75, fig. on 74 (1979); Balle in Fl. Cameroun 23: 51, t. 12/1–5 (1982), excl. subsp. *glaucoviridis*; Jones, Fl. Pl. Gambia: 49, t. 131 (1994)

15. **A. tanganyikae** (*Engl.*) *Polh. & Wiens* in Lebrun & Stork, Énum. Pl. Fl. Afr. Trop. 2: 165 (1992) & Mistletoes Afr.: 162 (1998). Type: Burundi, Kiriba Pass, *Scott Elliot* 8342 (B†, holo., BM!, K!, iso.)

Glabrous shrub; twigs slightly compressed, becoming terete. Leaves alternate to opposite; petiole 3–10 mm. long; lamina ovate to oblong-lanceolate or elliptic, 4–12 cm. long, 2–7 cm. wide, obtuse at the apex, broadly cuneate to rounded at the base, with 3–4 ascending nerves on each side. Umbels several, sessile in the axils, 2–3-flowered; sockets obliquely cupular; bract cupular with a small triangular limb, 2–4 mm. long. Receptacle 1.5–3 mm. long; calyx campanulate to tubular, 3–5 mm. long, slightly toothed, ciliolate, sometimes unilaterally split as the base of corolla swells. Corolla 4.5–6 cm. long, tube red, lobes yellow to orange over vents, with dark red to almost black tips; bud-tip not swollen; base of tube scarcely to slightly swollen at maturity, obovoid-fusiform, 8–9 mm. long, 3 mm. in diameter, the tube sometimes narrowed above for 5–10 mm. in diameter, then slightly flared; lobes erect, 8–10 mm. long, the upper two-thirds to three-quarters linear-lanceolate, slightly hardened inside. Filaments inflexed to once coiled, paler, hardened, the slightly expanded tip 1.5–2 mm. long; tooth 0.2–0.5 mm. long; anthers 2–3.5 mm. long. Style slightly broadened upwards, grooved, narrowed to a neck 2–3.5 mm. long; stigma capitate, 0.8 mm. across. Berry not seen.

TANZANIA. Ngara District: Ruvuvu R., Muwendo ferry, July 1953, *Eggeling* 6655!; Lushoto District: Kwamkoro–Sangarawe, Sept. 1955, *Semsei* 2327!; Morogoro District: W. slopes of Nguru Mts. above Maskati, 16 Mar. 1988, *Bidgood, Mwasumbi, Pócs & Vollesen* 452!
DISTR. T 1, 3, 6; Burundi
HAB. Riverine and montane forest, on various hosts, sometimes a pest of coffee; 900–2400 m.

SYN. *Loranthus tanganyikae* Engl. in E.J. 40: 531 (1908); Sprague in F.T.A. 6(1): 344 (1910); F.D.O.-A. 2: 172 (1932), pro parte; Balle in F.C.B. 1: 333 (1948)
    *L. sakarensis* Engl. in E.J. 40: 538 (1908); Sprague in F.T.A. 6(1): 367 (1910); F.D.O.-A. 2: 173 (1932); T.T.C.L.: 290 (1949). Type: Tanzania, Lushoto District, W. Usambara Mts., Sakare, *Engler* 943a (B!, holo., K!, fragment)
    *Tapinanthus sakarensis* (Engl.) Danser in Verh. K. Akad. Wet., sect. 2, 29(6): 119 (1933)
    *T. tanganyikae* (Engl.) Danser in Verh. K. Akad. Wet., sect. 2, 29(6): 120 (1933)
    *Agelanthus sakarensis* (Engl.) Polh. & Wiens in Lebrun & Stork, Énum. Pl. Fl. Afr. Trop. 2: 165 (1992)

NOTE. *Loranthus tanganyikae* and *L. sakarensis* were placed in different sections by Engler and Sprague, depending on whether the corolla is considered tubular or to have a slight but distinct swelling at the base. We considered at first that a distinction along these lines could be made between the western and eastern populations, but, with more material now available from the Nguru Mts., it does not really seem possible to maintain the distinction.

16. **A. irangensis** (*Engl.*) *Polh. & Wiens* in Lebrun & Stork, Énum. Pl. Fl. Afr. Trop. 2: 164 (1992) & Mistletoes Afr.: 163 (1998). Type: Tanzania, Kondoa District, W. of Irangi, *Stuhlmann* 4230 (B!, holo., K!, fragment)

Robust shrub; branchlets soon terete, subpersistently tomentellous. Leaves alternate to subopposite; petiole 2–4 mm. long; lamina coriaceous, green, ovate-elliptic, smaller ones sometimes obovate, 3–6.5 cm. long, 2–3 cm. wide, obtuse to rounded at the apex, obtuse to rounded and attenuate at the base, glabrous to thinly pubescent, with 3–6 pairs of lateral nerves, the second strongly ascending, sometimes inconspicuous. Umbels sessile, with 1–5 flowers in oblique sockets; bract cupular, 3–4.5 mm. long, puberulous. Receptacle 2–3 mm. long; calyx tubular, 7–8 mm. long, shortly toothed, ciliate, circumscissile near the base, free part remaining as a collar around the corolla-tube. Corolla 5.5–7 cm. long, tube red or orange-red, banded green at top, lobes yellow around vents, red-maroon at tip, without an apical swelling in bud; basal swelling slight, 6 mm. long, 3 mm. in diameter; lobes 14–17 mm. long, the upper two-thirds linear-oblanceolate. Filaments inflexed, the distal 2–3 mm. hardened and thickened; tooth 0.8–1 mm. long; anthers 3.5–5 mm. long. Style slender, slightly expanded opposite the filaments, narrowed to a neck 3.5–4 mm. long; stigma capitate, 0.8–1 mm. across. Berry not seen.

TANZANIA. Dodoma District: 18 km. E. of Itigi station, 11 Apr. 1964, *Greenway & Polhill* 11504!; Mpwapwa District: near Kibakwe village, 10 Apr. 1988, *Bidgood, Mwasumbi & Vollesen* 992!; Iringa District: Iringa–Ruaha National Park road, Iringa side of Ilusi bridge, 1986, *Congdon* 123!
DISTR. T 5, 7; not known elsewhere
HAB. Deciduous bushland and thicket, recorded on *Balanites* and *Boscia*; 1000–1650 m.

SYN. *Loranthus irangensis* Engl. in E.J. 20: 111 (1894) & P.O.A. C: 166, t. 13E–H (1895) & in E.J. 30: 302 (1901); Sprague in F.T.A. 6(1): 342 (1910); F.D.O.-A. 2: 172 (1932); T.T.C.L.: 290 (1949)
    *L. prittwitzii* Engl. in E.J. 40: 527 (1908); F.D.O-A. 2: 168 (1932). Type: Tanzania, Mbeya District, Utengule, *Prittwitz* 173 (B!, holo., K!, fragment)
    *Tapinanthus irangensis* (Engl.) Danser in Verh. K. Akad. Wet., sect. 2, 29(6): 113 (1933)

NOTE. A spectacular plant in flower but not often collected.

17. **A. validus** *Polh. & Wiens*, Mistletoes Afr.: 163 (1998). Type: Tanzania, Lushoto District, W. Usambara Mts., 3 km. S. of Malindi, Lukosi [Lukozi], *Drummond & Hemsley* 2685 (K!, holo., EA, SRGH!, iso.)

Forming large clumps, with longer branches pendent, glabrous; branchlets slightly compressed to angled, soon terete, moderately lenticellate. Leaves mostly alternate; petiole 3–15 mm. long; lamina slightly fleshy, dull green, ovate-elliptic, 5–10 cm. long, 2.5–7 cm. wide, obtuse to rounded at the apex, cuneate and attenuate at the base, with 4–5 pairs of lateral nerves, the second–third strongly ascending. Umbels sessile, with 2–3 flowers in sockets; bract cupular, 3–4 mm. long, slightly lacerate, ciliolate. Receptacle 2–3 mm. long; calyx tubular, 7–8 mm. long, slightly toothed, ciliate, circumscissile near the base, the freed upper part persistent around the corolla-tube. Corolla 4–5 cm. long, 3–4 mm. in diameter, red, yellow on lobes around vents, tip darker red, almost cylindrical in bud, only slightly swollen at the base; lobes 12 mm. long, linear-oblanceolate in the upper two-thirds. Filaments dark red, inflexed to shortly coiled, paler, hardened and thickened in the upper 1.5–2 mm.; tooth 0.7–1 mm. long; anthers 4 mm. long. Style slender, very slightly swollen opposite the filaments, with a neck 4 mm. long; stigma capitate, 0.8–1 mm. across. Young berries oblong-obovoid, warted.

TANZANIA. Lushoto District: Magamba Forest Reserve, 13 km. N. of Magamba, 19 Mar. 1972, *Wiens* 4591! & Shume-Magamba Forest Reserve, 9 July 1983, *Polhill & Lovett* 4973! & Manolo, 10 July 1983, *Polhill, Lovett & Ruffo* 4977!
DISTR. T 3; known only from the W. Usambara Mts.

HAB. In both wetter and drier montane forests, on various hosts including *Catha, Maytenus* and *Scolopia*, 1700–1950 m.

SYN. *Loranthus buchwaldii* Sprague in F.D.O.-A. 2: 178 (1932); T.T.C.L.: 292 (1949), *nomen*. Based on Tanzania, Lushoto District, W. Usambara Mts., Marimba, *Buchwald* 460! & Kwai, *Albers* 225!


18. **A. elegantulus** *(Engl.)* *Polh. & Wiens* in Lebrun & Stork, Énum. Pl. Fl. Afr. Trop. 2: 163 (1992) & in U.K.W.F., ed. 2: 156 (1994) & Mistletoes Afr.: 164, photo. 61, fig. 14C (1998). Type: Tanzania, Lushoto District, W. Usambara Mts., Kwa Mshusa, *Holst* 9071 (B!, holo., K!, P!, iso.)

Small shrub, spreading and rather compact, to 1 m.; branchlets slightly compressed, soon terete, glabrous except in axils, lenticellate to varying degrees. Leaves alternate; petiole 5–15 mm. long; lamina slightly fleshy, brownish to yellow-green, lanceolate to ovate-elliptic, 5–12 cm. long, 1.5–5 cm. wide, slightly acuminate to acute or obtuse at the apex, cuneate to rounded at the base, with 6–8 pairs of lateral nerves, the second–third ± strongly ascending. Umbels several, sessile, with 2–4 flowers in sockets; bract cupular, 1–2 mm. long, lacerate, usually with a subulate erose limb 1–3 mm. long. Receptacle 2 mm. long; calyx cupular, 2–3(–4 in fruit) mm. long, with distinct teeth 0.5–1 mm. long, ciliate, usually splitting unilaterally as the corolla expands. Corolla 3–4 cm. long, tube red, lobes yellow over vents, with red to red-brown tips; apical swelling of bud slight, 5 mm. long, apiculate; basal swelling oblong-ellipsoid to globose-ellipsoid, 5–6 mm. long, 2.5–3 mm. in diameter; lobes erect, linear-oblanceolate in upper two-thirds to three-quarters, apiculate, 7–8 mm. long. Filaments inflexed, yellow, the last 1 mm. hardened, paler, thickened; tooth 0.2–0.3 mm. long; anthers 2–3 mm. long. Style slender, slightly expanded and red opposite the filaments, narrowed to a neck 2–3 mm. long; stigma capitate, 0.6–0.8 mm. across. Berry bright red, narrowly obovoid, 6–8 mm. long, 4–5 mm. in diameter, warted especially on upper half; seed orange-yellow. Fig. 10/13–17 (p. 47).

KENYA. N. Nyeri District: midway between Nyeri and Mountain Lodge Hotel, 8 Dec. 1985, *Wiens* 6503!; Nairobi District: Karura Forest, 5 July 1981, *M.G. Gilbert & Ng'weno* 6276! & Westlands, 6 June 1975, *Gillett* 20821!

TANZANIA. Moshi District: Kilimanjaro, Bismarck Hill to Marangu, 1 Mar. 1934, *Greenway* 3882!; Lushoto District: W. Usambara Mts., saddle near Magamba Peak on Lushoto–Malindi road, 29 May 1953, *Drummond & Hemsley* 2806!; Morogoro District: Uluguru Mts., Lukwangule Plateau, Mar. 1955, *Semsei* 1967!

DISTR. **K** 4; **T** 2, 3, 6; not known elsewhere

HAB. Montane forest and upland dry evergreen forest, on a wide variety of hosts; 1500–2400 m.

SYN. *Loranthus elegantulus* Engl. in E.J. 20: 121 (1894) & P.O.A. C: 166, t. 17A (1895); Sprague in F.T.A. 6(1): 369 (1910); F.D.O.-A. 2: 174 (1932); T.T.C.L.: 289 (1949)
  *L. kilimandscharicus* Engl., P.O.A. C: 166 (1895); Sprague in F.T.A. 6(1): 343 (1910); F.D.O.-A. 2: 172 (1932); T.T.C.L.: 290 (1949). Type: Tanzania, Moshi District, Kilimanjaro, above Marangu, *Volkens* 1557 (B!, holo., BM!, iso., K!, fragment)
  *L. laciniatus* Engl., P.O.A. C: 166 (1895); Sprague in F.T.A. 6(1): 368 (1910); F.D.O.-A. 2: 173 (1932); T.T.C.L.: 290 (1949). Type: Tanzania, Moshi District, Kilimanjaro, above Kilema, *Volkens* 1818 (B!, holo., BM!, K!, iso.)
  *Odontella kilimandscharica* (Engl.) Tiegh. in Bull. Soc. Bot. Fr. 42: 260 (1895)
  *Loranthus holtzii* Engl. in E.J. 40: 526 (1908); Sprague in F.T.A. 6(1): 369 (1910); F.D.O.-A. 2: 174 (1932); T.T.C.L.: 289 (1949). Type: Tanzania, Lushoto District, W. Usambara Mts., Sakare, *Engler* 939a (B!, holo., K!, fragment)
  *L. keudelii* Engl. in E.J. 40: 538 (1908); Sprague in F.T.A. 6(1): 368 (1910); F.D.O.-A. 2: 174 (1932); T.T.C.L.: 290 (1949); U.K.W.F.: 334 (1974). Type: Tanzania, Lushoto [Wilhelmstal], *Keudel* in Herb. Amani 616g (B!, holo., K!, fragment)
  *L. rendelii* Engl. in F.D.O.-A. 2: 178 (1932); T.T.C.L.: 292 (1949), *nomen*. Based on Tanzania, Lushoto District, W. Usambara Mts., Mbalu, *Uhlig* 837 (B, K!)
  *Tapinanthus elegantulus* (Engl.) Danser in Verh. K. Akad. Wet., sect. 2, 29(6): 111 (1933)
  *T. holtzii* (Engl.) Danser in Verh. K. Akad. Wet., sect. 2, 29(6): 113 (1933)

*T. keudelii* (Engl.) Danser in Verh. K. Akad. Wet., sect. 2, 29(6): 114 (1933)
*T. kilimandscharicus* (Engl.) Danser in Verh. K. Akad. Wet., sect. 2, 29(6): 114 (1933)
*T. laciniatus* (Engl.) Danser in Verh. K. Akad. Wet., sect. 2, 29(6): 114 (1933)

NOTE. Similar to *A. sansibarensis*, with which it grows in the W. Usambara Mts., but altogether more delicate.

19. **A. bipartitus** *Polh. & Wiens* in Novon 7: 274, fig. 1F–H (1997) & Mistletoes Afr.: 165, fig. 14D (1998). Type: Tanzania, Njombe District, Madehani, *Stolz* 2475 (BR!, holo., BM!, FHO!, K!, P!, iso.)

Small shrub, spreading to 1 m.; branchlets scurfy at tips but soon thickened, grey, glabrescent and prominently lenticellate. Leaves alternate; petiole 1–5 mm. long; lamina thinly coriaceous, deep green, elliptic to oblong-elliptic, 4–8 cm. long, 2–3 cm. wide, obtuse to rounded at the apex, cuneate at the base, few-nerved, the second pair ascending almost to apex (sometimes scarcely apparent). Umbels 1–several, sessile, with 2–4 flowers in sockets; bract cupular, 2–3 mm. long, slightly lacerate, sparsely ciliolate, sometimes with a small limb. Receptacle 2 mm. long; calyx tubular, 3–5 mm. long, shortly toothed, ciliate, circumscissile near the base as the corolla swells, the detached part persistent on the corolla-tube. Corolla 3.2–4.2 cm. long, tube red, lobes yellow over vents, with darker red tip; apical swelling of bud slight, 5 mm. long; basal swelling globose-ellipsoid, 4 mm. long, 3–3.5 mm. in diameter; lobes erect, 7–8 mm. long, linear-oblanceolate in the upper two-thirds to three-quarters. Filaments inflexed, orange-yellow, the last 1 mm. hardened, paler, thickened; tooth 0.2–0.3 mm. long; anthers 2.5 mm. long. Style slender, red, slightly expanded opposite the filaments, narrowed to a neck 3 mm. long; stigma capitate, 0.7–0.8 mm. across. Berry not seen. Fig. 10/18 (p. 47).

TANZANIA. Iringa District: Mufindi, Lugoda Estate, 3 May 1986, *R. & D. Polhill* 5255!; Rungwe District: Kiwira R., Middle Fishing Camp, 26 Jan. 1963, *Richards* 17622!; Njombe District: Nyumbanyito, ± 13 km. WNW. of Njombe, 9 July 1956, *Milne-Redhead & Taylor* 11025!
DISTR. T 7; Malawi (Nyika Plateau)
HAB. Montane forest, on various hosts; 1800–2100 m.

SYN. [*Loranthus tanganyikae* sensu F.D.O.-A. 2: 172 (1932); T.T.C.L.: 290 (1949), pro parte quoad specim. *Stolz* 2475, *non* Engl.]

NOTE. Easily recognised by the calyx split around the base by the expanding corolla-base and retained as a dried ring around the narrow part of the corolla-tube. In other respects the species is similar to *A. elegantulus*, which occurs further north.

20. **A. sansibarensis** *(Engl.)* *Polh. & Wiens* in Lebrun & Stork, Énum. Pl. Fl. Afr. Trop. 2: 165 (1992), U.K.W.F., ed. 2: 156 (1994) & Mistletoes Afr.: 165 (1998). Type: Tanzania, Zanzibar I., *Stuhlmann* I. 772 (B!, holo.)

Shrub, spreading to 1 m., with a well-developed haustorial attachment, very rarely with subcortical runners, glabrous; branchlets slightly compressed to soon terete and densely lenticellate. Leaves mostly alternate, sometimes some ± opposite; petiole 3–20 mm. long; lamina coriaceous, dull green to yellow-green, often with a red flush, ovate to ovate-elliptic, 5–13 cm. long, 3–8(–11) cm. wide, subacute to rounded at the apex, rounded to cuneate and slightly attenuate at the base, several-nerved from near the base, other venation rather irregular. Umbels several, sessile, with 2–3 flowers in sockets; bract cupular, oblique, with a slight triangular limb and sometimes slightly lacerate, 2–3 mm. long. Receptacle 2–3 mm. long; calyx cupular, slightly toothed, 1–4 mm. long, ciliolate, split longitudinally into several segments as the corolla-base expands. Corolla 3–5 cm. long, tube red, sometimes green-banded at the top, lobes yellow over vents, with dark red to blackish tips; apical part of bud not swollen; basal swelling subglobular, 4–6 mm.

long, 3.5–6 mm. in diameter; lobes erect, slightly hardened in the upper half, pointed, 7–11 mm. long. Filaments inflexed, yellow-green then brownish, the last 1 mm. hardened, paler, thickened; tooth 0.3–0.5 mm. long; anthers 2–3 mm. long. Style slender, slightly expanded and red opposite the filaments, narrowed to a 3 mm. long neck; stigma capitate, 0.5–0.8 mm. diameter. Berry red, obovoid, 8–9 mm. long, 5–6 mm. in diameter, warty; seed red.

subsp. **sansibarensis**; Polh. & Wiens, Mistletoes Afr.: 166, photo. 62, fig. 14A (1998)

Calyx 1–2 mm. long, up to half as long as the basal swelling of the corolla-tube. Corolla-lobes 7–8 mm. long.

KENYA. Machakos District: junction of Ginia R. and Thika R., 9 May 1965, *Archer* 504!; Kwale District: Shimba Hills, Lango ya Mwagandi [Longomwagandi], 27 Feb. 1968, *Magogo & Glover* 194!; Lamu District: Mambasasa Forest Station, 29 Jan. 1958, *Verdcourt* 2127!

TANZANIA. Lushoto District: Bumbuli, 16 July 1983, *Polhill & Lovett* 5000!; Tanga District: coast near Bomalandani [Bomandani], 13 km. S. of Moa, 8 Aug. 1953, *Drummond & Hemsley* 3672!; Uzaramo District: Msasani beach, 30 May 1939, *Vaughan* 2814!; Zanzibar I., 27 km. towards Chwaka, 14 Feb. 1964, *Faulkner* 3341!

DISTR. **K** 4, 7; **T** 3, 6; **Z**; Somalia, Mozambique

HAB. Coastal bushland, mangrove swamps and forest, extending up rivers to montane and dry hilltop mist-forest, on a wide variety of hosts; 0–1950 m.

SYN. *Loranthus sansibarensis* Engl. in E.J. 20: 121 (1894) & P.O.A. C: 166 (1895); Sprague in F.T.A. 6(1): 366 (1910); Chiov., Fl. Somala 2: 387, fig. 222 (1932); F.D.O.-A. 2: 173 (1932), pro majore parte; T.T.C.L.: 290 (1949); U.O.P.Z.: 336, fig. (1949)
   *L. volkensii* Engl. in E.J. 20: 110 (1894) & P.O.A. C: 166 (1895). Types: Tanzania, Tanga, *Volkens* 183 (B!, syn., BM!, K!, isosyn.) & Lushoto/Tanga District, Bombo valley, near Fumbuli, *Holst* 2365 (B!, syn., COI!, isosyn., K!, fragment)
   *L. sadebeckii* Engl. in E.J. 20: 122, t. 3C (1894) & P.O.A. C: 166 (1895). Types: Tanzania, Tanga District, Doda, *Holst* 2970 (B!, syn., K!, isosyn.) & Bagamoyo, *Stuhlmann* 225 (B!, syn.) & Zanzibar I., *Hildebrandt* 1031 (B!, syn., BM!, isosyn.) & *Stuhlmann* I. 775 (B!, syn.)
   *L. crassissimus* Engl. in E.J. 20: 122 (1894), pro parte excl. *Stuhlmann* I. 1076, & P.O.A. C: 167, t. 14E–G (1895). Lectotype, chosen by Sprague (1910): Tanzania, Zanzibar I., Chuini [Tschueni], *Stuhlmann* I. 776 (B!, lecto.)
   *Odontella volkensii* (Engl.) Tiegh. in Bull. Soc. Bot. Fr. 42: 260 (1895)
   *Loranthus viridizonatus* Werth, Veg. Ins. Sansibar: 87 (1901). Type: Zanzibar I., *Werth* (B!, holo.)
   *L. tricolor* Peter, F.D.O.-A. 2: 173, Descr. 15, t. 15/2 (1932); T.T.C.L.: 290 (1949). Type: Tanzania, coast between Dar es Salaam and Bagamoyo, *Peter* 45081 (B, holo.)
   *Tapinanthus sansibarensis* (Engl.) Danser in Verh. K. Akad. Wet., sect. 2, 29(6): 119 (1933)
   *T. tricolor* (Peter) Danser in Rec. Trav. Bot. Néerl. 31: 224 (1934)

subsp. **montanus** *Polh. & Wiens*, Mistletoes Afr.: 166, fig. 14B (1998). Type: Kenya, N. Nyeri District, Nyeri, *Wiens* 4512 (K!, holo., EA, iso.)

Calyx 3–4 mm. long, nearly as long as the basal swelling of the corolla. Corolla-lobes 10–11 mm. long.

KENYA. Naivasha District: 19 km. W. of Ol Kalou, 16 Feb. 1972, *Wiens* 4564!; N. Nyeri District: Nyeri, 22 Jan. 1932, *V.G. van Someren* in *Napier* 1766!; S. Nyeri District: Aberdare Mts., Kabage, Apr. 1930, *Dale* in F.D. 2357!

TANZANIA. Moshi District: forest above Kilimanjaro Timbers, 25 June 1994, *Grimshaw* 94956!

DISTR. **K** 3, 4; **T** 2; known only from Mt. Kenya, Aberdares and Kilimanjaro

HAB. Montane forest; 1800–2500 m.

SYN. *Loranthus thorei* K. Krause in N.B.G.B. 8: 501, fig. 3 (1923). Lectotype, see Polh. & Wiens (1998): Kenya, N. Nyeri District, Nyeri, *R.E. & T.C.E. Fries* 2127 (B!, lecto., BR!, K!, UPS, isolecto.)
   *Tapinanthus thorei* (K. Krause) Danser in Verh. K. Akad. Wet., sect. 2, 29(6): 121 (1933)
   [*Loranthus sansibarensis* sensu Chiov., Racc. Bot. Miss. Consol. Kenya: 109 (1935), *non* Engl. sensu stricto]

21. **A. oehleri** *(Engl.) Polh. & Wiens* in Lebrun & Stork, Énum. Pl. Fl. Afr. Trop. 2: 164 (1992) & in U.K.W.F., ed. 2: 156 (1994) & Mistletoes Afr.: 166, photo. 63 (1998). Type: Tanzania, Masai District, Masai Steppe, *Oehler* 43 (B!, holo., K!, fragment)

Shrublet; stems often short, 5–50 cm. long, simple or with short flowering branches, ultimately more bushy, frequently connected by subcortical runners; branchlets slightly compressed, soon terete, glabrous, usually sparsely lenticellate. Leaves alternate, often little-developed at flowering time; petiole 1–10 mm. long; lamina thick, often reddish flushed, oblanceolate to obovate, 2.5–6(–9) cm. long, 1–3(–5) cm. wide, obtuse to rounded at the apex, attenuate at the base, 3–5-nerved from near the base (nerves sometimes obscure). Umbels several, sessile, with 2–3 flowers in sockets; bract cupular, oblique, with a slight triangular limb, 1–2 mm. long. Receptacle 1.5–2 mm. long; calyx cupular, 1.5–2.5 mm. long, split as the corolla-base expands. Corolla 3.5–4.5 cm. long, tube red, lobes yellow over vents, with darker red to maroon tip; apical part of bud not swollen; basal swelling ellipsoid-globose, 3.5–5 mm. long, 3–3.5 mm. in diameter; lobes erect, slightly hardened in upper half, pointed, 8–10 mm. long. Filaments inflexed, yellow, the last 1.5 mm. hardened, paler, thickened; tooth 0.3–0.5 mm. long; anthers 2.5–3.5 mm. long. Style slender, slightly expanded opposite the filaments, narrowed to a neck 3–4 mm. long; stigma capitate, 0.8 mm. across. Berry bright pink or red, narrowly obovoid, 7–8 mm. long, 4–6 mm. in diameter, coarsely tubercled especially above.

KENYA. Northern Frontier Province: SSW. of Mt. Nyiru, Kowop, 28 Oct. 1978, *M.G. Gilbert, Gachathi & Gatheri* 5170!; N. Nyeri District: Naibor road, 29 Dec. 1986, *Luke* 132!; Masai District: Nguruman Escarpment W. of Lake Magadi, 1 Aug. 1965, *Archer* 508!
TANZANIA. Mbulu District: Manyara National Park, Msasa, 20 Nov. 1963, *Greenway* in *E.A.* 12895! & main road to Mbulu on the W. Rift Wall from Mto wa Mbu, 25 Mar. 1968, *Greenway & Kanuri* 13238!; Pare District: 30 km. SE. of Same, 4 Apr. 1988, *Pócs* 88049E!
DISTR. **K** 1, 4, 6, 7; **T** 2, 3; S. Sudan
HAB. Deciduous and coastal bushland, on a variety of hosts, commonly *Commiphora* and *Euphorbia*, also smaller ± succulent shrubs, such as *Kleinia*, *Plectranthus* and *Ipomoea*; 20–2000 m.

SYN. *Loranthus oehleri* Engl. in E.J. 40: 526 (1908); Sprague in F.T.A. 6(1): 366 (1910); F.D.O.-A. 2: 173 (1932); T.T.C.L.: 290 (1949); U.K.W.F.: 334 (1974)
*L. obovatus* Peter, F.D.O.-A. 2: 168, Descr.: 13, t. 19/1 (1932); T.T.C.L.: 283 (1949), *nom. illegit.*, *non* Blume (1825). Type: Tanzania, Mbulu District, Iraku, Malimo, *Peter* 43772 (B, holo.)
*Tapinanthus oehleri* (Engl.) Danser in Verh. K. Akad. Wet., sect. 2, 29(6): 117 (1933); Blundell, Wild Fl. E. Afr.: 131, t. 515 (1987)
*T. obovatus* Danser in Rec. Trav. Bot. Néerl. 31: 224 (1934). Type as for *L. obovatus* Peter

NOTE. Closely related to *A. sansibarensis*, but occurring in drier habitats and generally easily recognised in the field by its smaller bushy habit, with frequent occurrence of subsidiary plants growing from subcortical runners, and more fleshy, often sparse, obovate leaves.

22. **A. fuelleborníi** *(Engl.) Polh. & Wiens* in Lebrun & Stork, Énum. Pl. Fl. Afr. Trop. 2: 163 (1992) & Mistletoes Afr.: 167, photo. 64 (1998). Type: Tanzania, Rungwe District [N. Nyasaland], without precise locality, *Goetze* 1899 (B!, holo., K!, fragment)

Well-branched shrub to 1 m., glabrous; branchlets slightly compressed, soon terete, ultimately densely lenticellate. Leaves usually alternate; petiole 5–20 mm. long; lamina coriaceous, bluish-green, elliptic to oblong-elliptic, sometimes slightly obovate, 4–13 cm. long, 2–7 cm. wide, obtuse to rounded at the apex, cuneate or rounded to attenuate at the base, with the few main nerves all ascending. Umbels (1–)2–3(–6)-flowered; peduncle 0–2(–3) mm. long; pedicels 0–2 mm. long in an oblique saucer-shaped socket; bract cupular, (1.5–)2–3 mm. long, ciliate, with a slight limb and sometimes puberulous on that side. Receptacle 2–2.5 mm. long; calyx cupular, 2–4 mm. long, shortly toothed, ciliolate, splitting unilaterally as the corolla swells. Corolla 3.5–5 cm. long, red, lobes yellow over vents, with darker red tips;

narrow above vents in bud; basal swelling ellipsoid to obovoid-globose, 5–6 mm. long, 3–4 mm. in diameter; tube constricted above for 4–6 mm.; lobes erect, 8–11 mm. long, the upper two-thirds linear-oblanceolate. Filaments inflexed, thickened in distal 1–2 mm.; tooth 0.3–0.5 mm. long; anthers 2.5–3 mm. long. Style slender, slightly expanded opposite the filaments, narrowed to a neck 3–4 mm. long; stigma capitate, 0.8–1 mm. across. Berry red, oblong-obovoid, 10 mm. long, 5–7 mm. in diameter, verrucose.

TANZANIA. Ufipa District: 3 km. from Sumbawanga on road to Mbisi, 26 Nov. 1994, *Goyder et al.* 3833!; Mbeya District: Mbeya Mt., 26 Jan. 1963, *Richards* 17541!; Songea District: R. Luhimba, ± 30 km. N. of Songea, 27 Jan. 1956, *Milne-Redhead & Taylor* 8561A!

DISTR. **T** 2, 4, 7, 8; Zambia, Malawi and Zimbabwe

HAB. Higher rainfall *Brachystegia* woodland and along rivers and lakes, on various hosts, including *Combretum, Protea, Dombeya, Flacourtia, Maytenus, Trichilia* and *Brachystegia*; 450–2100(–2500) m.

SYN. *Loranthus fuellebornii* Engl. in E.J. 30: 304 (1901); Sprague in F.T.A. 6(1): 365 (1910); F.D.O.-A. 2: 172 (1932); T.T.C.L.: 291 (1949)

[*L. sadebeckii* sensu Engl. in E.J. 30: 304 (1901), *non* Engl. (1894)]

*L. eylesii* Sprague in F.T.A. 6(1): 343 (1910) & in K.B. 1911: 146 (1911). Type: Zimbabwe, Bulawayo, *Eyles* 1194 (K!, holo., BM!, SRGH!, iso.)

*L. goetzei* Sprague in F.T.A. 6(1): 365 (1910) & in K.B. 1911: 149 (1911); F.D.O.-A. 2: 172 (1932), excl. specim. ex Usambara; T.T.C.L.: 291 (1949), pro parte. Type: Tanzania, Njombe District, Pikurugwe Mt., *Goetze* 989 (B!, holo., BR!, iso., K!, fragment)

*L. annulatus* Engl. & K. Krause in E.J. 51: 469 (1914); F.D.O.-A. 2: 172 (1932); T.T.C.L.: 289 (1949). Type: Tanzania, Rungwe District, Masoko, *Stolz* 1061 (B!, holo., K!, WAG!, iso.)

*Tapinanthus annulatus* (Engl. & K. Krause) Danser in Verh. K. Akad. Wet., sect. 2, 29(6): 107 (1933)

*T. eylesii* (Sprague) Danser in Verh. K. Akad. Wet., sect. 2, 29(6): 111 (1933)

*T. fuellebornii* (Engl.) Danser in Verh. K. Akad. Wet., sect. 2, 29(6): 112 (1933)

*T. goetzei* (Sprague) Danser in Verh. K. Akad. Wet., sect. 2, 29(6): 113 (1933)

[*L. brunneus* Engl. var. *durandii* sensu T.T.C.L.: 289 (1949), quoad specim. cit., *non* (Engl.) Sprague]

NOTE. Commonly confused with *A. sansibarensis*, but readily distinguished in the field by its glaucous leaves.

The type of *Loranthus goetzei*, on *Protea* at the exceptionally high altitude of 2500 m. in the Kipengere Range, with longer pedicels (2 mm.) and smaller bracts (1.5 mm.) than any other material, is not matched by any modern material, but is nevertheless probably conspecific.

23. **A. nyasicus** (*Baker & Sprague*) *Polh. & Wiens* in Lebrun & Stork, Énum. Pl. Fl. Afr. Trop. 2: 164 (1992) & Mistletoes Afr.: 170 (1998). Types: Malawi, Namadzi [Namasi], *Cameron* 8 & without locality, *Buchanan* 1090 (K!, syn., B, fragments!)

Bushy shrub to 1.5 m., with relatively short lateral branches, glabrous except very occasionally on the inflorescences and even more rarely on youngest twigs; branchlets compressed at first, pruinose, soon terete, rather stout, brown to grey, with late emergence of lenticels. Leaves mostly subopposite, more alternate on long shoots, with smaller, more rounded, shorter petioled leaves clustered at the base of lateral shoots; petiole 0.5–2(–3) cm. long; lamina glaucous (?always), coriaceous, lanceolate to ovate-elliptic or oblong-elliptic, 3–12 cm. long, 2–6 cm. wide, bluntly pointed at the apex, cuneate or rounded at the base, with 3–6 pairs of lateral nerves, the lower ones more ascending. Umbels or heads (4–)6–16-flowered; peduncle 5–12 mm. long and pedicels 5–8 mm. long, normally glabrous, occasionally puberulous; bract ovate-triangular, pointed to truncate, 1.5–2 mm. long. Receptacle 2–3 mm. long; calyx 0.3–0.8 mm. long, with broad shallow teeth. Corolla 4–6 cm. long, orange to pale flame-red, with yellow band around vents, bud-tip darker and slightly ribbed, glabrous or very rarely with a few small scattered hairs; basal swelling ovoid-ellipsoid to ovoid-globose, 5–9 mm. long, 4–6 mm. in diameter, with the tube constricted for 2–11 mm. above; lobes erect, 9–11 mm. long; upper expanded part linear-elliptic,

5–7 mm. long, 1–1.5 mm. in diameter, hardened inside, keeled outside. Stamens inflexed; tooth 0.3–0.5 mm. long; anthers 2.5–4 mm. long. Style with a neck 2.5–4 mm. long; stigma capitate, 0.8–1 mm. across. Berry red, obovoid, 10–12 mm. long, 7–8 mm. in diameter.

TANZANIA. Ufipa District: Sumbawanga, 24 Nov. 1949, *Bullock* 1933!; Morogoro, near forest, 9 Dec. 1932, *Wallace* 540!; Iringa District: Dabaga Highlands, 31 km. E. of Iheme on the Dabaga road, 12 Feb. 1962, *Polhill & Paulo* 1445!
DISTR. T 4, 6, 7; SE. Zaire, Zambia, Malawi, N. Mozambique, S. Zimbabwe and E. Botswana
HAB. Edges of montane forest, in riverine forest and generally higher rainfall *Brachystegia* woodland, on various hosts, most commonly *Ficus, Bridelia* and various legumes; 1000–1900 m.

SYN. *Loranthus nyasicus* Baker & Sprague in F.T.A. 6(1): 371 (Dec. 1910) & in K.B. 1911: 150 (1911); F.D.O.-A. 2: 174 (1932); T.T.C.L.: 291 (1949)
    *L. stolzii* Engl. & K. Krause in E.J. 45: 291 (Dec. 1910); Sprague in F.T.A. 6(1): 1033 (1913); Engl. in V.E. 3(1): 102, fig. 67 (1915); F.D.O.-A. 2: 174 (1932), excl. specim. ex Pare; Engl. & K. Krause in E. & P. Pf., ed. 2, 16B: 166, fig. 84 (1935); T.T.C.L.: 291 (1949), pro parte. Type: Tanzania, Rungwe District, Bundali Mts., *Stolz* 113 (B!, holo., BM!, K!, MO!, US!, iso.)
    *L. luteostriatus* Engl. & K. Krause in E.J. 51: 464 (1914); F.D.O.-A. 2: 174 (1932); T.T.C.L.: 291 (1949). Type: Tanzania, Rungwe District, Bundali, *Stolz* 1797 (B!, holo., BM!, BR!, K!, MO!, P!, iso.)
    *L. lateritiostriatus* Engl. & K. Krause in E.J. 51: 465 (1914); F.D.O.-A. 2: 174 (1932); T.T.C.L.: 291 (1949). Type: Tanzania, Rungwe District, Ulambya [Bulambya], *Stolz* 1615 (B, holo., BM!, BR!, K!, MO!, P!, iso.)
    *Tapinanthus lateritiostriatus* (Engl. & K. Krause) Danser in Verh. K. Akad. Wet., sect. 2, 29(6): 115 (1933).
    *T. luteostriatus* (Engl. & K. Krause) Danser in Verh. K. Akad. Wet., sect. 2, 29(6): 115 (1933)
    *T. nyasicus* (Baker & Sprague) Danser in Verh. K. Akad. Wet., sect. 2, 29(6): 117 (1933).
    *T. stolzii* (Engl. & K. Krause) Danser in Verh. K. Akad. Wet., sect. 2, 29(6): 120 (1933)

NOTE. The forest-edge forms are probably less glaucous and sometimes redder flowered than those from *Brachystegia* woodland.

24. **A. entebbensis** (*Sprague*) Polh. *& Wiens* in Lebrun & Stork, Énum. Pl. Fl. Afr. Trop. 2: 163 (1992) & Mistletoes Afr.: 172 (1998). Type: Uganda, Mengo District, Entebbe, *Rutter* (K!, holo., BM!, iso.)

Stems spreading to 1 m. or more; branchlets slightly flattened, pubescent with short spreading hairs, the longer ones branched, red-brown and shed first, gradually glabrescent. Leaves alternate to opposite; petiole 5–15 mm. long; lamina thinly coriaceous, slightly glossy, narrowly lanceolate to ovate-elliptic, 6–14 cm. long, 2–5 cm. wide, acute to obtuse at the apex, cuneate at the base, soon glabrous except sometimes near the base; lateral nerves 4–8 pairs, mostly obliquely ascending (uppermost more spreading). Umbels 6–12-flowered; peduncle 3–5 mm. long and pedicels 4–5 mm. long, both densely short spreading puberulous; bract-limb ovate-triangular, 2 mm. long. Receptacle 2.5 mm. long, puberulous; calyx 0.7–1 mm. long, toothed. Corolla (4.5–)5–5.5 cm. long, red, banded yellow around vents, with darker red tip, minutely puberulous on bud-head and thinly pubescent with short branched hairs on tube; apical swelling of bud oblong-ellipsoid, ribbed; basal swelling ovoid-ellipsoid, 6–8 mm. long, 4–5 mm. in diameter, narrowed above for 5–7 mm.; lobes erect, 9–11 mm. long, the upper part linear-elliptic, 5–6 mm. long, 0.8–1 mm. in diameter, hardened inside, slightly keeled outside. Stamens inflexed; tooth 0.2–0.5 mm. long; anthers 2–3 mm. long. Style with a neck 2–3 mm. long; stigma capitate, 0.8 mm. across. Berry not seen.

UGANDA. Busoga District: Jinja, Nov. 1958, *Tweedie* 1726!; Mbale District: Elgon, Sipi and Sabei, 11 Apr. 1927, *Snowden* 1053!; Masaka District: 5 km. N. of Lake Nabugabo, 25 Apr. 1971, *Lye* 5990!
KENYA. Elgon, 2100 m., Nov. 1932, *Jack* 372!

DISTR. U 3, 4; **K** 3/5; not known elsewhere
HAB. At forest edges, commonly on *Ficus*, and in wooded grassland, commonly on *Combretum*; 1100–2100 m.

SYN. *Loranthus entebbensis* Sprague in K.B. 1913: 181 (1913)
   *Tapinanthus entebbensis* (Sprague) Danser in Verh. K. Akad. Wet., sect. 2, 29(6): 111 (1933)

NOTE. This species may prove to be no more than a form of *A. keilii* when more material is available for comparison. It is also closely related to *A. musozensis*, which is more hairy.

25. **A. keilii** (*Engl. & K. Krause*) *Polh. & Wiens* in Lebrun & Stork, Énum. Pl. Fl. Afr. Trop. 2: 164 (1992) & Mistletoes Afr.: 173 (1998). Type: Burundi, Bujumbura [Usumbura], near Kamata, *Keil* 212 (B!, holo., K!, fragment)

Shrub flowering mostly on relatively short lateral branches; branchlets slightly flattened, puberulous with short spreading hairs, soon glabrescent. Leaves subopposite; petiole 5–15 mm. long; lamina thinly coriaceous, somewhat glaucous, lanceolate to ovate-elliptic or oblong-elliptic, 5–11.5 cm. long, 1–4 cm. wide, acute to obtuse at the apex, cuneate at the base, glabrous except for a few scattered hairs near the base, with 4–6 pairs of lateral nerves, the lower ones ± strongly ascending. Umbels 8–12-flowered; peduncle 7–10 mm. long, covered with stiff spreading hairs; pedicels 6–8 mm. long, similarly hairy; bract-limb ovate-triangular, 1–2 mm. long. Receptacle 1.8–2 mm. long, minutely puberulous to subglabrous; calyx 0.3–0.5 mm. long, shortly and broadly toothed. Corolla 4–4.5 cm. long, dull yellow to red with mauve tip, minutely puberulous on the lobes, subglabrous to sparsely puberulous on the tube (more conspicuous on constriction); apex of bud narrowly oblong-elliptic, 5–6 mm. long, 2.5 mm. in diameter, slightly ribbed; basal swelling ovoid-globose, 6 mm. long, 4 mm. in diameter, narrowed for 2–3 mm. above; lobes erect, 9 mm. long, the upper expanded part linear-elliptic, 5–5.5 mm. long, 1 mm. wide, hardened inside, keeled outside. Stamens inflexed; tooth 0.5 mm. long; anthers 2.5–3 mm. long. Style with a neck 2.5–3 mm. long; stigma capitate, 0.7 mm. across. Berry not seen.

TANZANIA. Buha District: near Mohorro, *Peter* 37679
DISTR. **T** 4; S. Rwanda and Burundi
HAB. On *Ficus*; 1350–1450 m.

SYN. *Loranthus keilii* Engl. & K. Krause in E.J. 43: 409 (1909); Sprague in F.T.A. 6(1): 376 (1910); F.D.O.-A. 2: 175 (1932); Balle in F.C.B. 1: 341 (1948); T.T.C.L.: 287 (1949)
   *Tapinanthus keilii* (Engl. & K. Krause) Danser in Verh. K. Akad. Wet., sect. 2, 29(6): 114 (1933)
   *Loranthus becquetii* Balle in B.J.B.B. 17: 241 (1944). Type: Burundi, Kafumbe [Kafiumbe], *Becquet* 126 (BR!, holo.)
   *Tapinanthus prunifolius* (Harv.) Tiegh. subsp. *keilii* (Engl. & K. Krause) Balle in Troupin, Fl. Rwanda 1: 189 (1978); Troupin, Fl. Pl. Lign. Rwanda: 379 (1982), *nom. invalid., comb. non rite publ.*

NOTE. The single specimen recorded from Tanzania has not been seen. When the complex is better known the material attributed to *A. entebbensis* might be included in this species. The small differences are noted in the key.

26. **A. molleri** (*Engl.*) *Polh. & Wiens* in Lebrun & Stork, Énum. Pl. Fl. Afr. Trop. 2: 164 (1992) & Mistletoes Afr.: 173, photo. 67 (1998). Type: Angola, Pungo Andongo, Cuanza R., Calembe I., *Welwitsch* 4845 (B!, holo., BM!, COI!, K!, LISU!, P!, iso.)

Much-branched shrub to 1 m., flowering mostly on relatively short lateral shoots; branchlets slightly flattened at first, densely covered with short spreading hairs, longer ones branched, red-brown and shed first. Leaves mostly opposite to subopposite, but those at the base of lateral shoots crowded, smaller, shorter stalked, more obovate and with fewer more ascending nerves; petiole 5–15 mm. long; lamina thinly coriaceous, dull green to slightly glaucous, lanceolate to elliptic-lanceolate or ovate-elliptic, 4–13 cm. long, 0.5–3 cm. wide, narrowed or slightly acuminate to an acute or obtuse apex, cuneate to rounded at the base, pubescent to glabrous except near the base, with 4–6 pairs of lateral nerves, the lower ones ascending. Umbels (1–)2–8-flowered; peduncle (0–)1.5–3 mm. long, densely spreading hairy; pedicels 2–5 mm. long, densely hairy; bract-limb ovate-triangular, 1.5–2 mm. long. Receptacle 1.5 mm. long, hairy to almost glabrous; calyx flared, 1–1.5 mm. long, split by the expanding corolla, ciliate. Corolla 3.8–4.2 cm. long, greenish white, orange-red around vents, with greenish tip flushed dull red outside, usually dull red inside tube along the lines of filament-attachments, densely covered with mostly branched hairs 1–2 mm. long; apical swelling of bud slight, 5 mm. long, 1.5–2 mm. in diameter; basal swelling ellipsoid-globose, 4–5 mm. long, 3–4.5 mm. in diameter, narrowed for 3–4 mm. above; lobes erect, 8–9 mm. long, the upper slightly expanded part linear-oblanceolate, 5–7.5 mm. long, 0.6–0.8 mm. wide, slightly hardened inside especially above. Stamens red, the upper third paler, thickened and hardened, with a very small tooth; anthers 1.8–2.5 mm. long. Style often flushed red opposite the filaments, with a neck 2–2.5 mm. long; stigma capitate, 0.5–0.7 mm. across. Berry oblong-obovoid, 8 mm. long, 4 mm. in diameter, crowned by persistent calyx, pubescent.

TANZANIA. Ufipa District: Tatanda Mission, 23 Feb. 1994, *Bidgood, Mbago & Vollesen* 2415!; Iringa District: Madibira, Nyamakuyu, 20 Jan. 1985, *Congdon* 32!; Songea District: near R. Luhimba, 6 May 1956, *Milne-Redhead & Taylor* 10104!
DISTR. **T** 2, 4, 7, 8; S. Zaire, Angola, Zambia, Malawi and Zimbabwe
HAB. Wooded grassland, generally on *Combretum*, sometimes on *Acacia* and elsewhere recorded on *Ficus*; 950–1700 m.

SYN. *Loranthus molleri* Engl. in E.J. 20: 120 (1894); Sprague in F.T.A. 6(1): 364 (1910); Balle in F.C.B. 1: 338 (1948)
  *Tapinanthus molleri* (Engl.) Danser in Verh. K. Akad. Wet., sect. 2, 29(6): 116 (1933)

27. **A. songeensis** *Polh. & Wiens*, Mistletoes Afr.: 175 (1998). Type: Tanzania, Songea District, 8 km. S. of Gumbiro, *Milne-Redhead & Taylor* 10148 (K!, holo., B!, BR!, EA, LISC!, SRGH!, iso.)

Large shrub, spreading and pendulous, flowering mostly on relatively short lateral branches; branchlets stout, slightly flattened, minutely velvety, the slightly longer hairs reddish brown, branched and falling first, gradually glabrescent. Leaves opposite and subopposite; petiole 1–2 cm. long; lamina deep green, somewhat blue-green beneath, coriaceous, ovate to ovate-elliptic, 7–17 cm. long, 3–12 cm. wide, bluntly pointed at the apex, basally cuneate to rounded, scabridulous with minute branched hairs especially on midvein and beneath, glabrescent, with 4–9 pairs of main lateral nerves, the lower ones obliquely ascending. Umbels 6–8-flowered; peduncle 4–5 mm. long; pedicels 3–4 mm. long, red-brown velvety; bract-limb ovate-triangular to elliptic-concave, 3–4 mm. long. Receptacle 2.5 mm. long, red-brown velutinous; calyx cupular, 0.8–1 mm. long, subentire. Corolla 4.5–5 cm. long, reddish on bulbous base, then vivid orange-vermilion, banded orange-yellow at the base of the lobes with red tips, velutinous with minute red-brown branched hairs; apex of bud shortly oblong-ellipsoid, 5 mm. long, 3.5 mm. in diameter, slightly angled and with slight keels raised into small subapical prominences; basal swelling ovoid-ellipsoid, 8–10 mm. long, 4.5–5 mm. in diameter, narrowed for 5–6 mm. above; lobes erect, 11 mm. long, the upper expanded part 5.5–6 mm. long, 1.2–1.5 mm. wide, hardened inside, slightly crested outside. Stamens pink shading to orange-yellow, inflexed; tooth 0.5–0.7 mm. long; anthers 3–3.5 mm. long. Style red opposite the filaments, with a neck 3–3.5 mm. long above; stigma 1 mm. across. Berry not seen.

TANZANIA. Songea District: 8 km. S. of Gumbiro, 10 May 1956, *Milne-Redhead & Taylor* 10148! & Mbinga township, 29 Apr. 1991, *Ruffo & Kisena* 3362! & towards bottom of Mbamba Bay escarpment by road near Ngamba Mission, 6 Apr. 1956, *Milne-Redhead & Taylor* 9540!

DISTR. **T** 8; Malawi

HAB. Deciduous woodland and wooded grassland, on *Piliostigma* and elsewhere on *Combretum*; 550–950 m.

NOTE. A very handsome species liable to confusion with *Phragmanthera*, but the detailed structure of the hairs, bracts, stamens, etc. is quite different.


28. **A. musozensis** (*Rendle*) *Polh. & Wiens* in Lebrun & Stork, Énum. Pl. Fl. Afr. Trop. 2: 164 (1992) & in U.K.W.F., ed. 2: 156 (1994) & Mistletoes Afr.: 175, photo. 68 (1998). Type: Uganda, Masaka District, Misozi [Musozi], *Bagshawe* 134 (BM!, holo.)

Spreading and pendent shrub to 2 m., flowering mostly on relatively short lateral branches; branchlets slightly flattened, velvety with very short spreading branched hairs. Leaves subopposite; petiole 5–10 mm. long; lamina slightly fleshy, linear-lanceolate to ovate or oblong-elliptic, 5–10 cm. long, 1.5–8 cm. wide, acute to obtuse at the apex, cuneate to rounded or cordate at the base, puberulous (more densely along midrib), ultimately glabrescent, with 4–6 pairs of spreading-ascending main lateral nerves. Umbels 6–12-flowered; peduncle and pedicels 2–5 mm. long, velvety; bract-limb ovate-triangular, 2–3 mm. long. Receptacle 3 mm. long, densely hairy; calyx cupular, 1–1.5 mm. long, slightly toothed. Corolla 4.5–5.5 cm. long, pale orange-yellow, banded yellow around vents, sometimes greenish at the base, tip turning reddish, densely covered with branched hairs, the longer ones wispy and ± reddish; apex of buds oblong-ellipsoid to narrowly obovoid, 6–7 mm. long, 2.5–3 mm. in diameter, slightly ribbed below; basal swelling ovoid-ellipsoid, 7–8 mm. long, 4 mm. in diameter, narrowed for 4–5 mm. above; lobes erect, 9–11 mm. long, the upper expanded part linear-elliptic, 5.5–6.5 mm. long, 1–1.2 mm. wide, hardened inside. Stamens red, inflexed; tooth 0.3–0.5 mm. long; anthers 3–4 mm. long. Style red opposite vents, with a neck 3–4 mm. above; stigma capitate, 0.8–1 mm. across. Berry red, obovoid, 10 mm. long, 8 mm. in diameter, pubescent.

UGANDA. Masaka District: Misozi [Musozi], Jan. 1904, *Bagshawe* 134!

KENYA. S. Kavirondo District: 13 km. Homa Bay to Mbita Point, 5 km. from Ogondo, 4 Nov. 1980, *M.G. & C.I. Gilbert* 5977!; Kericho District: Chepalungu Forest, Feb. 1931, *Gardner* in *F.D.* 2514!; Masai District: 2.5 km. W. of Ngore Ngore and Keekorok road junction on road to Lolgorien, 20 Dec. 1985, *D. & C. Wiens* 6509!

TANZANIA. Maswa District: Serengeti Plains, Moru Kopjes, 31 Dec. 1971, *Greenway & Myles Turner* 14954!; Mbulu District: Mwembe, 5 Jan. 1962, *Polhill & Paulo* 1053!; E. Mpwapwa, 20 June 1934, *Hornby* 592!

DISTR. **U** 4; **K** 4–6; **T** 1, 2, 5; Zaire and Rwanda

HAB. Forest edges to wooded grassland or bushland, often on rock outcrops, exclusively on *Ficus*; 1100–1900 m.

SYN. *Loranthus musozensis* Rendle in J.L.S. 37: 204 (1905); Sprague in F.T.A. 6(1): 382 (1910)
    *L. erythrotrichus* K. Krause in N.B.G.B. 8: 503, fig. 2A–D (1923). Type: Kenya, Meru, *R.E. & T.C.E. Fries* 1774 (B!, holo., BR!, K!, iso.)
    *Tapinanthus erythrotrichus* (K. Krause) Danser in Verh. K. Akad. Wet., sect. 2, 29(6): 111 (1933)
    *T. musozensis* (Rendle) Danser in Verh. K. Akad. Wet., sect. 2, 29(6): 116 (1933)
    *Loranthus brideliae* Balle var. *oblongifolius* Balle in B.J.B.B. 17: 243 (1944). Type: Zaire, between Beni and Ruwenzori, *Lebrun* 4398 (BR!, holo.)
    *L. blantyreanus* Engl. var. *oblongifolius* (Balle) Balle in F.P.N.A. 1: 104 (1948)
    *Tapinanthus prunifolius* (Harv.) Tiegh. subsp. *musozensis* (Rendle) Balle in Troupin, Fl. Rwanda 1: 189, fig. 38/1 (1978); Troupin, Pl. Lign. Rwanda: 379, fig. 132.1 (1982), *nom. invalid., comb. non rite publ.*


29. **A. uhehensis** (*Engl.*) *Polh. & Wiens* in Lebrun & Stork, Énum. Pl. Fl. Afr. Trop. 2: 165 (1992) & Mistletoes Afr.: 177 (1998). Type: Tanzania, Iringa District, near Dabaga, *Goetze* 641 (B†, holo., K!, fragment)

Stems spreading to 1.5 m. long; branchlets scurfy-tomentellous with small much-branched red-brown hairs. Leaves opposite to alternate; petiole 1–3 cm. long; lamina dull mid-green, thinly coriaceous, linear-lanceolate to narrowly elliptic-oblong, 4–13 cm. long, 0.5–5 cm. wide, acute to obtuse at the apex, cuneate at the base, scurfy-pubescent, soon glabrescent, with 8–10 pairs of spreading lateral nerves. Umbels (or heads) 6–16-flowered; peduncle 5–8 mm. long, scurfy; pedicels 3–5 mm. long, similarly hairy; bract-limb linear to oblong-lanceolate, eventually produced into a slight foliaceous limb, 4–12 mm. long. Receptacle 1.5–2 mm. long, red-brown scurfy; calyx 0.8–1 mm. long, toothed. Corolla 3.5–4.5 cm. long, orange, becoming redder inside, scurfy puberulous with short much-branched red-brown hairs, apex of buds shortly oblong-ellipsoid, 4 mm. long, 2 mm. in diameter, angular; basal swelling ellipsoid, 5–6 mm. long, 3–4 mm. in diameter, narrowed for 3–4 mm. above; lobes erect, 7–9 mm. long, the upper expanded part oblanceolate to linear-elliptic, 3–4 mm. long, 1 mm. wide, hardened inside, slightly keeled outside. Stamens red, inflexed; tooth 0.2 mm. long; anthers 2–3 mm. long. Style red opposite vents, narrowed to a 2–3 mm. long neck; stigma capitate, 0.8 mm. across. Berry red, obovoid-ellipsoid, 8–10 mm. long, 6–7 mm. in diameter, scurfy; seed dark yellow.

TANZANIA. Iringa District: Mufindi, Livalonge Forest Division, 26 Mar. 1985, *Congdon* 39! & Brooke Bond Tea Estates, Kibao, 5 Jan. 1986, *D. & C. Wiens & Congdon* 6545! & Kivere Estate, 4 May 1986, *R. Polhill, D. Polhill & Congdon* 5265!
DISTR. **T** 7, 8 (see note); not known elsewhere
HAB. Forest edges, commonly on *Bridelia*; 1800–2000 m.

SYN. *Loranthus uhehensis* Engl. in E.J. 28: 382 (1900); Sprague in F.T.A. 6(1): 378 (1910); F.D.O.-A. 2: 175 (1932); T.T.C.L.: 288 (1949)
    *Tapinanthus uhehensis* (Engl.) Danser in Verh. K. Akad. Wet., sect. 2, 29(6): 121 (1933)

NOTE. Easily confused with species of *Phragmanthera*, and liable to be overlooked, but the hairs are irregularly much-branched, not with whorls of branches, the anthers are not chambered and the berries are red.
    Seen by Congdon in Songea District, Kiteza Forest Reserve, but no specimen collected.

30. **A. combreticola** (*Lebrun & L. Touss.*) *Polh. & Wiens* in Lebrun & Stork, Énum. Pl. Fl. Afr. Trop. 2: 163 (1992) & Mistletoes Afr.: 177 (1998). Type: Rwanda, Mushushu [Muchuchu] to Urwita [Murwita], *Lebrun* 9669 (BR!, holo.)

Much-branched shrub; branchlets stout, shortly hairy with branched red-brown hairs. Leaves opposite or subopposite; petiole 6–16 mm. long; lamina coriaceous, ovate-elliptic, rarely obovate, 5–9 cm. long, 3–5.5 cm. wide, obtuse or rounded at the apex, cuneate at the base, rather densely puberulous with branched hairs when young, glabrescent except on the nerves, with 6–8 pairs of main lateral nerves, the second–third rather strongly ascending. Umbels mostly 4-flowered; peduncle 1–2 mm. long, glabrous to puberulous; pedicels 3–4 mm. long, densely puberulous; bract-limb ovate-triangular, 2.5 mm. long. Receptacle 2 mm. long, puberulous; calyx 0.5–0.7 mm. long, truncate. Corolla 5.5–6 cm. long, lemon-yellow, with orange-red tip in bud, puberulous (hairs 0.5 mm.); apical swelling of buds truncate, 4 mm. across, 5-winged; basal swelling ellipsoid, 6–8 mm. long; lobes erect, 1 cm. long, upper half expanded, hardened inside, strongly keeled outside. Stamens inflexed; tooth 0.7 mm. long; anthers 3–4 mm. long. Style with a neck 3 mm. long; stigma capitate, 1 mm. across. Berry not seen.

TANZANIA. Mpanda District: Kampisa R., 14 Feb. 1996, *Congdon* 459!
DISTR. **T** 4; Rwanda
HAB. Wooded grassland, on *Combretum* and *Terminalia*; 1600 m.

SYN. *Loranthus combreticola* Lebrun & L. Touss. in Lebrun, Expl. Parc Nat. Kagera: 45 (1948); Balle in F.C.B. 1: 342 (1948), as '*combreticolus*'

*Tapinanthus prunifolius* (Harv.) Tiegh. subsp. *combreticola* (Lebrun & L. Touss.) Balle in Troupin, Fl. Rwanda 1. 188 (1978), as '*combreticolus*', Troupin, Fl. Pl. Lign. Rwanda. 379 (1982), as '*combreticolus*', *nom. invalid., comb. non rite publ.*

31. **A. scassellatii** (*Chiov.*) *Polh. & Wiens* in Lebrun & Stork, Énum. Pl. Fl. Afr. Trop. 2: 165 (1992) & in U.K.W.F., ed. 2: 157 (1994) & Mistletoes Afr.: 178, photo. 70, fig. 15A (1998). Type: Somalia, Giamama, *Scassellatii & Mazocchi* 111 (FT, holo.)

Spreading shrub to 1 m., with relatively short lateral flowering branches; branchlets slightly compressed, soon terete and densely lenticellate, subglabrous to densely puberulous on youngest parts. Leaves mostly subopposite; petiole 5–15(–25) mm. long; lamina yellow-green turning dark green, slightly fleshy, lanceolate to oblong-elliptic, rarely broadly elliptic, 4–9(–12) cm. long, 1.5–4(–5) cm. wide, bluntly pointed, ± slightly acuminate at the apex, cuneate or rounded at the base, glabrous, with few main lateral nerves, the lower ones ascending. Umbels 4–12-flowered; peduncle 1–5 mm. long, puberulous; pedicels 3–6 mm. long, puberulous; bract triangular-lanceolate, pointed or truncate, 1.5–3 mm. long. Receptacle 1.5–2 mm. long, glabrous to puberulous; calyx 0.5 mm. long, with short broad teeth, ciliate. Corolla 3.8–4.2 cm. long, orange to reddish, yellow or yellow-green at the base, yellow or orange-yellow banded around vents, tip of bud yellow or green, ± darkening red, puberulous especially on apical swelling, very sparsely to distinctly puberulous elsewhere; apex in bud oblong to obovoid, rounded to subtruncate, strongly ribbed, 4 mm. long, 2.5–3 mm. in diameter; basal swelling ovoid-globose, 4–6 mm. long, 3.5–4.5 mm. in diameter, narrowed for 4–8 mm. above; lobes erect, 8–9 mm. long; upper expanded part oblanceolate, 3.5–4.5 mm. long, 1 mm. wide, hardened inside, strongly keeled outside. Stamens orange or red, inflexed to involute; tooth 0.3 mm. long; anthers 2–3 mm. long. Style red opposite vents, narrowed to a paler 2.5–3 mm. long neck; stigma 0.8 mm. across. Berry red. Fig. 10/19–23 (p. 47).

KENYA. Northern Frontier Province: 42 km. SE. of Malka Mari, 22 Jan. 1972, *Bally & Smith* 14938!; Embu District: 5 km. W. of Ishiara, 18 Jan. 1972, *S.A. Robertson* 1739!; Kilifi District: N. end of Mangea Hill, 29 Mar. 1990, *Luke & S.A. Robertson* 2231!
TANZANIA. Tanga District: Manza–Sabutuni road, 8 km. S. of Moa, 11 Aug. 1953, *Drummond & Hemsley* 3701!; Bagamoyo District: Kidomole, Apr. 1964, *Semsei* 3834!; Kilwa District: Selous Game Reserve, Mcharaya area, 13 Feb. 1978, *Vollesen* in *M.R.C.* 4936!; Zanzibar I., Chwaka, 27 Dec. 1930, *Vaughan* 1782!
DISTR. K 1, 4, 7; T 3, 6, 8; Z; S. Somalia
HAB. Coastal and deciduous bushland, extending up rivers to hilltop mist-forest and montane forest, on various hosts; 0–2300 m.

SYN. *Loranthus celtidifolius* Engl. in E.J. 20: 123, t. 3D (1894) & P.O.A. C: 167 (1895); Sprague in F.T.A. 6(1): 372 (1910); F.D.O.-A. 2: 174 (1932), *nom. illegit., non* Schult. & Schult.f. (1829). Type: Tanzania, Uzaramo District, Dar es Salaam, *Hildebrandt* 1225 (B!, holo., BM!, K!, P!, iso.)
   *L. aurantiacus* Engl. in E.J. 20: 124 (1894) & P.O.A. C: 167, t. 14A–D (1895); Sprague in F.T.A. 6(1): 372 (1910); F.D.O.-A. 2: 175 (1932); U.K.W.F.: 330 (1974), *nom. illegit., non* Hook. (1848). Types: Tanzania, Tanga District, Amboni, *Holst* 2890 (B!, syn., K!, P!, isosyn.) & Bagamoyo, *Stuhlmann* 137 (B!, syn.)
   *L. scassellatii* Chiov., Result. Sci. Miss. Stef.-Paoli, Coll. Bot. 1: 218 (1916)
   *L. amblyphyllus* Chiov., Fl. Somala 2: 384, fig. 219 (1932). Type: Somalia, Kismayu [Chisimaio], *Senni* 137 (FT, holo.)
   *Tapinanthus aurantiacus* Danser in Verh. K. Akad. Wet., sect. 2, 29(6): 108 (1933); Vollesen in Opera Bot. 59: 64 (1980). Types as for *L. aurantiacus* Engl.
   *T. celtidifolius* Danser in Verh. K. Akad. Wet., sect. 2, 29(6): 110 (1933). Type as for *L. celtidifolius* Engl.
   *T. scassellatii* (Chiov.) Danser in Verh. K. Akad. Wet., sect. 2, 29(6): 119 (1933)
   *Loranthus aurantiaciflorus* Brenan in K.B. 4: 94 (1949) & T.T.C.L.: 287 (1949). Types as for *L. aurantiacus* Engl.
   *L. celtidiformis* Brenan in K.B. 4: 94 (1949) & T.T.C.L.: 287 (1949). Type as for *L. celtidifolius* Engl.

NOTE. Specimens from the Tanga area, including one of the syntypes of *L. aurantiacus*, have unusually short obovoid heads to the flower-buds.

32. **A. igneus** (*Danser*) *Polh. & Wiens* in Lebrun & Stork, Énum. Pl. Fl. Afr. Trop. 2: 164 (1992) & Mistletoes Afr.: 178 (1998). Type: Tanzania, Lindi District, Cheminda to Bakari-Rondo, *Braun in Herb. Amani* 1169 (B!, holo., EA, iso., K!, fragment)

Well-branched shrub with relatively short lateral branches; branchlets compressed at first with scurfy puberulence only on first 1–2 internodes, soon terete, brown turning grey, fissured and prominently lenticellate. Leaves opposite to subopposite; petiole 5–12 mm. long; lamina dull green, thinly coriaceous, linear-lanceolate to lanceolate or elliptic-lanceolate, 4–12 cm. long, 1–3.5 cm. wide, pointed at the apex, glabrous, with up to 8 pairs of main lateral nerves, lower ones strongly ascending. Umbels 6–8-flowered; peduncle 2–3 mm. long, puberulous; pedicels 4–5 mm. long, puberulous; bract-limb ovate-triangular, pointed to truncate, 1–1.5 mm. long. Receptacle 1.5–2 mm. long, glabrous to scurfy puberulous; calyx 0.5 mm. long, broadly toothed. Corolla 3.5–3.8 cm. long, deep orange to pinkish red or fiery red, yellow around vents, darker red-brown at tip, subglabrous with minute scurfy puberulence on the lobe-tips and ± some small hairs on the tube; apex of buds oblong, subacute, 3–4 mm. long, 2 mm. in diameter; basal swelling ovoid-globular, 3.5–5 mm. long, 3–4 mm. in diameter, narrowly constricted for 3–4 mm. above; lobes erect, 6–8 mm. long, the upper hardened part linear-lanceolate, 3–4 mm. long, hardened inside, rounded outside. Stamens inflexed; tooth minute; anthers 1.5–2.5 mm. long. Style with a neck 1.5–2.5 mm. long; stigma capitate, 0.8 mm. across. Berry not seen.

TANZANIA. Lindi District: Rondo Plateau, 7 Apr. 1998, *Congdon* 509! & Cheminda to Bakari-Rondo, 12 June 1906, *Braun in Herb. Amani* 1169!
DISTR. T ?6 (see note), 8; N. Mozambique south to the Zambezi
HAB. *Brachystegia* woodland, but in Mozambique in riverine forest and coastal bushland, recorded on *Grewia* and *Combretum*; 800 m.

SYN. *Loranthus igneus* Sprague in K.B. 1912: 232 (1912) & in F.T.A. 6(1): 1034 (1913); F.D.O.-A. 2: 175 (1932); T.T.C.L.: 287 (1949), *nom. illegit., non* (Scheff.) Benth. & Hook.f. (1880)
     *Tapinanthus igneus* Danser in Verh. K. Akad. Wet., sect. 2, 29(6): 113 (1933). Based on *L. igneus* Sprague

NOTE. Closely related to *A. scassellatii* and *A. transvaalensis* (Sprague) Polh. & Wiens, but easily recognised by the lack of distinct keels to the back of the corolla-lobes.
     Peter (1932) also records the species from Mafia I., but the specimen, *Braun in Herb. Amani* 3615, has not been seen.

33. **A. atrocoronatus** *Polh. & Wiens*, Mistletoes Afr.: 179, photo. 71 (1998). Type: Tanzania, Iringa, *Polhill & Paulo* 1694 (K!, holo., B!, BR!, EA!, P!, SRGH!, iso.)

Shrub to 1 m., flowering mostly on relatively short lateral branches; branchlets glabrous to minutely puberulous with short stiff spreading hairs. Leaves mostly subopposite, but smaller, more obovate and more crowded at the base of lateral shoots; petiole 8–15 mm. long; lamina lanceolate to oblong-elliptic, 5–10 cm. long, 1.5–4 cm. wide, acute or obtuse at the apex, cuneate or rounded at the base, glabrous, with 6–8 pairs of lateral nerves, the lower ones sometimes strongly ascending. Umbels 6–8-flowered; peduncle 3–6 mm. long, puberulous; pedicels 2–5 mm. long, puberulous; bract-limb ovate-triangular, 1.5–2 mm. long, pointed or truncate. Receptacle 1.5–2 mm. long, thinly puberulous; calyx 0.3–0.5 mm. long, broadly toothed. Corolla 3.5–3.8 cm. long, tube red, lobes yellow with purple-brown tips, puberulous on head of bud (lobes), sparsely hairy on tube (the longest hairs on the constriction and basal swelling reddish and deflexed); apical swelling of bud cylindrical, 4–4.5 mm. long, 2 mm. in diameter, with flattened thumb-like appendages from near the top of each lobe giving the form of a crown; basal swelling ovoid-globose, 3.5–4 mm. across, narrowed for 3–5 mm. above; lobes erect, 7 mm. long, the upper expanded part lanceolate, 4–4.5 mm. long, 1 mm. wide, hardened inside and slightly thickened towards the base, spurred outside. Stamens pink, inflexed; tooth 0.5 mm. long; anthers 2 mm. long. Style purplish opposite vents, with a 2 mm. long neck above; stigma capitate, 0.8 mm. across. Berry not seen.

TANZANIA. Iringa District: Dabaga Highlands, Kilolo, 10 Feb. 1962, *Polhill & Paulo* 1420! & 29 km. SE. of Tanzam highway on W. loop road to Dabaga, 29 Jan. 1986, *Wiens & Frank* 6598! & Mufindi, Lugoda golf course, 26 Feb. 1987, *Lovett* 1604!
DISTR. **T** 7; known only from the Iringa plateau
HAB. Edges of forest and deciduous woodland, on various hosts; 1650–1900 m.

NOTE. The strikingly crowned corolla-buds are an unusual feature for species of this genus which have vents, compared to the heads typical of *Tapinanthus*. M.G. Gilbert has noticed in Ethiopia that *A. heteromorphus* (A. Rich.) Polh. & Wiens (a close ally of *A. musozensis*) tends to have flowers opened by birds squeezing the heads rather than slitting the corolla open through the vents, and similar signals may be given by this species.

34. **A. pungu** (*De Wild.*) Polh. *& Wiens* in Lebrun & Stork, Énum. Pl. Fl. Afr. Trop. 2: 165 (1992) & Mistletoes Afr.: 180, photo. 72, fig. 16D (1998). Type: Zaire, Shaba, Lukafu, *Verdick* 388 (BR!, holo., K!, fragment)

Rounded shrub to 1.5 m., more pendulous with age, flowering on shoots of varying length; branchlets compressed at first, soon terete, densely pubescent with short spreading hairs. Leaves mostly opposite or subopposite; petiole 5–20 mm. long; lamina grey-green to markedly blue-green, slightly fleshy, linear-lanceolate to ovate-elliptic or broadly elliptic, 4–10 cm. long, 1–5.5 cm. wide, bluntly pointed to rounded at the apex, cuneate to rounded or cordate at the base, glabrous, with ascending nerves, either pinnate or lower 2–3 pairs stronger. Umbels or heads 8–24-flowered; peduncle and pedicels 2–5 mm. long, spreading-puberulous; bract-limb ovate-triangular, 1–1.5 mm. long, pointed or truncate. Receptacle 1.2–2 mm. long, puberulous, sometimes glabrescent; calyx 0.2–0.5 mm. long, shortly to distinctly toothed. Corolla 3.5–4(–4.5) cm. long, yellow to yellow-green, banded yellow or orange on lobes with tips green turning dull red outside, pubescent with longer spreading hairs on tube above basal swelling, somewhat glabrescent overall; apical swelling of bud fusiform-conical, sometimes slightly ribbed; basal swelling ovoid-globular, 3–5 mm. long, 3.5–4 mm. in diameter, narrowly constricted above for 2–4 mm. in diameter; lobes erect, 5–8 mm. long, the upper expanded part linear-lanceolate, 3–5 mm. long, 0.8–1 mm. wide, hardened inside, rarely keeled. Stamens red, inflexed; tooth 0.1–0.2 mm. long; anthers 2–3 mm. long. Style red opposite vents, with a neck 2–3 mm. long; stigma capitate, 0.8–1 mm. across. Berry cherry-red, obovoid, pubescent.

TANZANIA. Ufipa District: Tatanda, 19 Nov. 1994, *Goyder et al.* 3752!; Njombe District: 23 km. S. of Iyayi on Chalowe road, 12 Jan. 1986, *D. & C. Wiens & Congdon* 6566!; Songea District: 5.5 km. E. of Songea, 13 Feb. 1956, *Milne-Redhead & Taylor* 8684!
DISTR. **T** 4, 7, 8; Zaire (Shaba), Zambia, Malawi, Mozambique, Zimbabwe, Namibia (Caprivi Strip) and northernmost South Africa
HAB. *Brachystegia* woodland and wooded grassland, but extending to forest edges and down major drainage systems to mixed deciduous bushland, on a wide variety of hosts, but closer examination will often show that it is epiparasitic on other species of Loranthaceae; 700–2100 m.

SYN. *Loranthus pungu* De Wild. in Ann. Mus. Congo, Bot., sér. 4, 1: 175, t. 40/1–5 (1903); Sprague in F.T.A. 6(1): 377 (1910)
        *L. ceciliae* N.E. Br in K.B. 1906: 168 (1906), as '*cecilae*'; Sprague in F.T.A. 6(1): 373 (1910), as '*cecilae*'. Type: Zimbabwe, Bulawayo, *Cecil* 96 (K!, holo.)
        *L. blantyreanus* Engl. in E.J. 40: 537 (1908); Sprague in F.T.A. 6(1): 375 (1910). Type: Malawi, Blantyre, *Buchanan* in Herb. *J.M. Wood* 6983 (B!, holo., BM!, K!, iso.)

*L. ceciliae* N.E. Br. var. *buchananii* Sprague in F.T.A. 6(1): 374 (1910) & in K.B. 1911: 151
(1911). Type: Malawi, Blantyre, *Buchanan* 133 (K!, holo., MAL!, iso.)

*L. carsonii* Baker & Sprague in F.T.A. 6(1): 376 (1910) & in K.B. 1911: 151 (1911); F.D.O.-
A. 2: 175 (1932). Type: Zambia, Mbala District, Fwambo, *Carson* (K!, holo., B!, fragment)

*L. glaucescens* Engl. & K. Krause in E.J. 51: 465 (1914); F.D.O.-A. 2: 175 (1932); T.T.C.L.: 287
(1949). Type: Tanzania, Rungwe District, Masoko, *Stolz* 1062 (B!, holo., G!, K!, LD!,
WAG!, iso.)

*L. luteiflorus* Engl. & K. Krause in E.J. 51: 466 (1914); F.D.O.-A. 2: 175 (1932); T.T.C.L.: 288
(1949). Type: Tanzania, Rungwe District, Mulinda Forest, *Stolz* 1714 (B, holo., BM!, BR!,
EA!, FHO!, K!, P!, iso.)

*L. pungu* De Wild. var. *angustifolius* De Wild., Contr. Fl. Katanga: 53 (1921). Type: Zaire,
Shaba, Kapiri valley, *Hombié* 1941 (BR!, holo., K!, iso.)

*Tapinanthus blantyreanus* (Engl.) Danser in Verh. K. Akad. Wet., sect. 2, 29(6): 108 (1933)

*T. carsonii* (Baker & Sprague) Danser in Verh. K. Akad. Wet., sect. 2, 29(6): 109 (1933);
Wiens & Tölken in F.S.A. 10: 9, fig. 3/2 (1979)

*T. ceciliae* (N.E. Br.) Danser in Verh. K. Akad. Wet., sect. 2, 29(6): 110 (1933), as '*cecilae*';
Wiens & Tölken in F.S.A. 10: 9, fig. 3/1 (1979)

*T. glaucescens* (Engl. & K. Krause) Danser in Verh. K. Akad. Wet., sect. 2, 29(6): 112 (1933)

*T. luteiflorus* (Engl. & K. Krause) Danser in Verh. K. Akad. Wet., sect. 2, 29(6): 115 (1933)

*T. pungu* (De Wild.) Danser in Verh. K. Akad. Wet., sect. 2, 29(6): 118 (1933)

[*Loranthus dichrous* sensu Balle in F.C.B. 1: 339 (1948), pro parte, quoad specim. cit., *non*
Engl.]

NOTE. The species is generally easily distinguished from its nearest allies by the short spreading
indumentum on the branchlets, the grey-green or more generally glaucous leaves, the
sparsely hairy corolla with longer hairs on the tube, and the corolla-lobes without or with only
slight keels. The leaves are very variable in shape and this accounts for a considerable part of
the synonymy.

35. **A. schweinfurthii** (*Engl.*) *Polh. & Wiens* in Lebrun & Stork, Énum. Pl. Fl. Trop.
Afr. 2: 165 (1992) & Mistletoes Afr.: 182 (1998). Types: Sudan, Equatoria, Seriba
Ghattas, *Schweinfurth* 1381 (B!, syn., K!, P!, isosyn.) & 1425A (B!, syn.) & Tanzania,
Mwanza District, Bukumbuli [Bukombi], *Stuhlmann* 820 (B!, syn.)

Well-branched shrub; branchlets slightly compressed, densely pubescent with
short stiff spreading hairs; older branchlets glabrescent, grey, cracked,
inconspicuously lenticellate. Leaves opposite and subopposite; petiole 5–15 mm.
long; lamina slightly fleshy, dull green, lanceolate to oblong-elliptic, ovate or broadly
elliptic, 5–9 cm. long, 3–5.5 cm. wide, pointed to rounded at the apex, cuneate to
rounded at the base, glabrous except on the prominently raised lower part of midrib
beneath, with 6–8 pairs of lateral nerves, the lower ones more ascending. Umbels
8–16-flowered; peduncle and pedicels 2–4 mm. long, densely pubescent; bract-limb
ovate-triangular, pointed or truncate, 1.5–2.5 mm. long. Receptacle 1.5–2 mm. long,
glabrous to densely puberulous; calyx 0.3–0.8 mm. long, shallowly toothed. Corolla
3.5–4.2 cm. long, greenish yellow to orange-red, tipped brownish red, pubescent with
small irregularly arranged hairs; apex of buds narrowly ellipsoid, 4–5 mm. long,
2–2.5 mm. in diameter, pentagonal; basal swelling ovoid-ellipsoid, 5–7 mm. long,
3.5–4 mm. in diameter, constricted for 3–4 mm. above; lobes erect, 8–9 mm. long,
the upper dilated part linear-elliptic, (4–)5–6 mm. long, 1–1.5 mm. wide, slightly
keeled on lower part outside. Stamens inflexed; tooth 0.3–0.5 mm. long; anthers
2.5–3 mm. long. Style with a neck 2.5–3 mm. long; stigma capitate, 1 mm. across.
Berry not seen.

UGANDA. W. Nile District: Mt. Otzi, 7 June 1936, *A.S. Thomas* 1982!; Acholi District: Chua*,
*Eggeling* 2323!; Teso District: Katakwi [Katikwe], July 1930, *Liebenberg* 265!
TANZANIA. Mwanza District: Bukumbuli [Bukombi], Nov. 1890, *Stuhlmann* 820!
DISTR. U 1, 3; T 1; Central African Republic, Zaire and S. Sudan

* Data from label, field notebook gives Bunyoro District, Budongo, Nov. 1935.

Hab. Riverine forest, wooded grassland and bushland, on various hosts including *Combretum* and *Ficus*, 1100–1500 m.

Syn. *Loranthus schweinfurthii* Engl. in E.J. 20: 124 (1894) & P.O.A. C: 167 (1895); Sprague in F.T.A. 6(1): 375 (1910); F.D.O.-A. 2: 175 (1932), excl. distrib. Cameroon; T.T.C.L.: 288 (1949); F.P.S. 2: 293 (1952)
*Tapinanthus schweinfurthii* (Engl.) Danser in Verh. K. Akad. Wet., sect. 2, 29(6): 120 (1933)


## 9. **TAPINANTHUS**

(Blume) Rchb., Repert. Herb. Nom. Gen. Pl.: 73 (1841); Tiegh. in Bull. Soc. Bot. Fr. 42: 267 (1895); Polh. & Wiens, Mistletoes Afr.: 183 (1998)

*Loranthus* L. sect. *Tapinanthus* Blume, Fl. Javae, Lorantheae: 15 (1830)
*L.* sect. *Tapinanthus* Blume group *Constrictiflori* Engl. in E.J. 20: 113 (1894), pro parte
*Acrostephanus* Tiegh. in Bull. Soc. Bot. Fr. 42: 267 (1895)
*Loranthus* L. sect. *Constrictiflori* (Engl.) Sprague in F.T.A. 6(1): 268 (1910), *nom. superfl.*

Shrubs, mostly 0.5–2 m. from a single haustorial connection; twigs slightly compressed at first, soon terete, glabrous to puberulous or velvety with small simple hairs. Leaves mostly opposite, sessile to petiolate; lateral nerves spreading, usually some of the lower ones more strongly ascending. Flowers 2–16 in umbels, 5-merous; umbels in axils or clustered at older nodes, shortly pedunculate, perulate; bracts saucer-shaped to cupular with a small triangular limb. Calyx saucer-shaped to cupular or occasionally shortly tubular, almost entire, sometimes split by the expanding corolla. Corolla-tube much longer than the lobes, opening with a V-split, usually pink to purplish, sometimes spotted white; bud-head greenish to white, darkening at maturity, commonly angled, ribbed, winged or appendaged; basal swelling well formed; lobes attenuate to spathulate, reflexed or elsewhere sometimes erect. Filaments inserted near top of corolla-tube, short, linear, inflexed to inrolled; tooth distinct, 0.5–1 mm. long; anthers 4-thecous, short, (1.2–)2–3.5 mm. long, with a broad connective slightly produced at the apex. Style swollen opposite the filaments, constricted above; stigma small, obovoid to capitate, 0.4–1 mm. across. Berry oblong-ellipsoid to globose, often somewhat urceolate to the persistent calyx, smooth to papillose, usually ripening red.

30 species in tropical and southern Africa, one species extending to N. Yemen, virtually absent from the Somali-Masai and Afromontane regions, but in a wide range of associations elsewhere.

The genus is restricted here to the species in Sprague's *Loranthus* section *Constrictiflori*. These species are characterised by an apical swelling of the corolla-bud that darkens and often secretes nectar at the sutures at maturity as a signal to birds to peck it and release a targeted spray of pollen from the relatively small anthers, the corolla-lobes generally reflexing.

As a consequence of the bud-head acting as a signal to pollinators, it has become modified to a considerable extent with wings and appendages. Unfortunately much of the variation occurs within natural polymorphic species, notably *T. dependens* and *T. globiferus*, and limits its usefulness in distinguishing species. The modifications, however, often show some ecogeographic pattern.

1. Corolla hairy · · · · · · · · · · · · · · · · · · · · · · · · · · · · · · · · · · · · · · · · · · · · 2
   Corolla glabrous · · · · · · · · · · · · · · · · · · · · · · · · · · · · · · · · · · · · · · · · · · 3
2. Leaves with a well-developed petiole, the blade cuneate
   to rounded or slightly cordate · · · · · · · · · · · · · · · · · ·   2. *T. buvumae*
   Leaves subsessile, cordate to amplexicaul · · · · · · · · · · · ·   5. *T. erianthus*
3. Umbels (4–)6–8-flowered; receptacle barrel-shaped,
   longer than shallow calyx · · · · · · · · · · · · · · · · · · · ·   6. *T. globiferus*
   Umbels 2–4-flowered; receptacle obconic, ± as long as
   the calyx at anthesis · · · · · · · · · · · · · · · · · · · · · · · · · · · · · · · · · · · · · · · 4

4. Leaves acuminate · · · · · · · · · · · · · · · · · · · · · · · · · · ·          1. *T. constrictiflorus*
   Leaves acute to rounded at tip, but not acuminate · · · · · · · · · · · · · · · · · · 5
5. Calyx 2–4 mm. long; corolla-tube pink to crimson · · · · ·          3. *T. dependens*
   Calyx 1–1.5 mm. long; corolla-tube usually dull red at
   least along filament-lines · · · · · · · · · · · · · · · · · · · · ·          4. *T. forbesii*

1. **T. constrictiflorus** (*Engl.*) *Danser* in Verh. K. Akad. Wet., sect. 2, 29(6): 110 (1933); Balle & Troupin in Troupin, Fl. Rwanda 1: 189 (1978); Troupin, Fl. Pl. Lign. Rwanda: 378 (1982); Polh. & Wiens in U.K.W.F., ed. 2: 157 (1994) & Mistletoes Afr.: 189, photo 76 (1998). Lectotype, chosen by Sprague (1910): Tanzania, Bukoba, *Stuhlmann 3974* (B!, lecto., K!, fragment)

Plant glabrous; twigs slightly compressed at first. Leaves mostly opposite; petiole 5–15(–20) mm. long; lamina ± coriaceous, green to slightly glaucous, lanceolate to ovate, 4.5–15 cm. long, 1.2–8 cm. wide, acuminate at the apex, rounded to cordate, rarely broadly cuneate, at the base, with 8–12 pairs of lateral nerves, lower 2–3 strongly ascending. Umbels (2–)4-flowered; peduncle 2–4 mm. long; pedicels 0.5–3 mm. long from well-developed sockets; bract saucer-shaped, with a small triangular limb, 1–2 mm. long. Receptacle obconic, 1–2 mm. long; calyx shortly cupular, 1–2 mm. long, ciliolate. Corolla-tube 3.2–4 cm. long, pink to crimson, sometimes darker along filament-lines, head of buds green turning blue-grey to purple; bud-heads oblong-ellipsoid, obtuse to truncate, angled to ribbed or narrowly winged, sometimes with bosses or triangular points up to 1 mm. long overtopping the apex, 3.5–4 mm. long, 2–2.5 mm. in diameter; basal swelling 3.5–6 mm. long, 2.5–4 mm. in diameter, with the tube constricted for 3–8 mm. above; lobes reflexed, 6–7 mm. long, slightly corrugated on claw, expanded tip narrowly elliptic, 3.5–4 mm. long, 1–1.5 mm. wide (excluding margins). Stamens with tooth 0.5–0.7 mm. long; anthers 2–2.5 mm. long. Style with a neck 2–2.5 mm. long; stigma capitate, 0.7–0.8 mm. across. Berry urceolate, 10 mm. long, 6 mm. in diameter.

UGANDA. Toro District: Ruwenzori, Nyinabitaba [Nyabitaba], Jan. 1951, *Osmaston 3738*!; Kigezi District: Kayonza Forest, June 1957, *V.G.L. van Someren* 1!; Mengo District: Banda, May 1915, *Dummer* 2463!

KENYA. Nandi Hills township, 31 Mar., 1961, *Archer* 237!; S. Kavirondo District: Lake Victoria, Rusinga I., May 1941, *Opiko* in *Bally* 1473! & 16 km. S. of Mbita, Roo valley, 2 June 1978, *Plaizier* 1350!

TANZANIA. Bukoba District: Minziro, Feb. 1993, *Kielland*! & Maruku, *Panayotis* 91! & Lubale, Nov. 1931, *Haarer* 2367!

DISTR. U 2, 4; K 3, 5; T 1; Gabon, Zaire, Rwanda, Burundi, S. Sudan and N. Angola

HAB. Evergreen forest, on various hosts including plantation crops; 1100–2300 m.

SYN. *Loranthus constrictiflorus* Engl. in E.J. 20: 119, t. 3B (1894), excl. specim. ex Angola, & P.O.A. C: 166 (1895); Sprague in F.T.A. 6(1): 351 (1910); F.D.O.-A. 2: 172 (1932); Balle in F.C.B. 1: 344 (1948); F.P.N.A. 1: 98 (1948); T.T.C.L.: 291 (1949); U.K.W.F.: 332, fig. on 335 (1974)
   *L. syringifolius* Engl. in E.J. 20: 115 (1894) & P.O.A. C: 166 (1895); Sprague in F.T.A. 6(1): 350 (1910); F.D.O.-A. 2: 172 (1932), *nom. illegit.*, *non* Schult. (1829). Type: Uganda, Toro District, Ruwenzori, *Stuhlmann* 2470 (B!, holo., K!, fragment)
   *Acrostephanus syringifolius* Tiegh. in Bull. Soc. Bot. Fr. 42: 268 (1895). Type as for *L. syringifolius* Engl.
   *Loranthus pittospori* Rendle in J.L.S. 37: 205 (1905), as '*pittosporae*'. Type: Uganda, Kigezi District, Rukiga [Ruchigga], *Bagshawe* 445 (BM!, holo.)
   *L. constrictiflorus* Engl. var. *karaguensis* Sprague in F.T.A. 6(1): 352 (1910) & in K.B. 1911: 148 (1911); F.D.O.-A. 2: 172 (1932); T.T.C.L.: 291 (1949). Types: Tanzania, Bukoba District, Karagwe, Mtagata, *Stuhlmann* 3190 & Bukoba, *Stuhlmann* 4019A (B!, syn., K, fragments!)
   *Tapinanthus syringifolius* (Tiegh.) Danser in Verh. K. Akad. Wet., sect. 2, 29(6): 120 (1933)

NOTE. This species is somewhat variable, particularly with respect to the shape of the head of the buds which has been used as a key character to discriminate between groups of species.

The head is generally obtuse with only slight ribs or wings, but the lobes can have bosses which in extreme forms make a distinct crown of points around the sunken apex. The extreme form, resembling the shape found in *T. buvumae* from the same general area, is consistent around the north and east sides of Lake Victoria, and found to a lesser extent in NE. Zaire, but var. *karaguensis* refers to forms with small bosses and links to normal forms on the Tanzanian side of Lake Victoria and sporadically in Zaire.

2. **T. buvumae** (*Rendle*) *Danser* in Verh. K. Akad. Wet., sect. 2, 29(6): 109 (1933); Polh. & Wiens in U.K.W.F., ed. 2: 157, t. 55 (1994) & Mistletoes Afr.: 191, photo. 77 (1998). Type: Uganda, Lake Victoria, Buvuma I., *Bagshawe* 628 (BM!, holo.)

Stems to 1.5 m. long; branchlets velutinous, gradually glabrescent. Leaves mostly opposite; petiole 5–15(–30) mm. long; lamina ± coriaceous, green or yellow-green, lanceolate to ovate-elliptic, 6–14 cm. long, 1.5–8 cm. wide, generally slightly acuminate at the apex, cuneate to rounded or cordate at the base, glabrous, with 4–8 pairs of lateral curved-ascending nerves, the second pair sometimes stronger and ascending to upper half. Umbels in axils and crowded at older nodes, (2–)4-flowered, puberulous overall; peduncle 3–5 mm. long; pedicels from slight sockets, 1.5–3 mm. long; bract saucer-shaped, with a small triangular limb, 2–3 mm. long. Receptacle 1.5–2 mm. long; calyx also 1.5–2 mm. long. Corolla-tube 3.3–3.8 cm. long, puberulous, dull pink to reddish purple, with whitish or greenish spots on upper half, the head of buds grey-green with darker angles and the lobes green inside; bud-heads oblong, angled to narrowly winged, overtopped by an erect or spreading crown of 5 ovate-triangular points 1–2 mm. long, altogether 4–5 mm. long, 2.5–3 mm. in diameter; basal swelling obovoid-ellipsoid, 4–6 mm. long, 3–4 mm. in diameter, with the tube constricted for 3–6 mm. above; lobes 5–6 mm. long, reflexed, transversely corrugated inside, the upper expanded part oblong-elliptic, 3–3.5 mm. long, 1–1.2 mm. wide. Stamens green; tooth 0.5–0.7 mm. long; anthers 2–2.5 mm. long. Style green above, with a neck 2–2.5 mm. long; stigma capitate, 0.8 mm. across. Berry orange turning red, urceolate, 10 mm. long, 6 mm. in diameter.

UGANDA. Bunyoro District: Busingiro Forest Station, Nov. 1939, *Eggeling* 3838!; Teso District: Kaberamaido, Prison Farm, 6 Sept. 1951, *Stephens* 13!; Mengo District: Lwamatuka, 15 Sept. 1954, *Langdale-Brown* 1295!
KENYA. N. Kavirondo District: SW. slopes of Elgon, Jan. 1934, *Tweedie* 118!
TANZANIA. Mwanza District: Ukerewe I., Neuwied, 15 July 1904, *Conrads* 1507! & Mwanza, 28 Mar. 1952, *Tanner* 605! & 5 km. from Kikongo ferry on Geita road, 17 July 1960, *Verdcourt* 2884!
DISTR. U 2–4; K 5; T 1; E. Zaire
HAB. Wooded grassland, on various hosts, commonly *Combretum* or *Grewia*; 750–1200 m.

SYN. *Loranthus buvumae* Rendle in J.L.S. 37: 207 (1905); Sprague in F.T.A. 6(1): 357 (1910); Balle in F.C.B. 1: 352 (1948); F.P.N.A. 1: 100, t. 8 (1948)
    *L. conradsii* Sprague in F.D.O.-A. 2: 178 (1932); T.T.C.L.: 292 (1949), *nomen*. Based on Tanzania, Mwanza District, Ukerewe I., Neuwied, *Conrads* 1507 (BM!, K!)

NOTE. Very similar to *Tapinanthus constrictiflorus*, but occurring in drier habitats and recognisable by the pubescence and consistently crowned corolla-buds.

3. **T. dependens** (*Engl.*) *Danser* in Verh. K. Akad. Wet., sect. 2, 29(6): 110 (1933); Polh. & Wiens, Mistletoes Afr.: 192, photo. 78, fig. 17C (1998). Type: Angola, Pungo Andongo, Mutollo and Candumba, *Welwitsch* 4851 (B!, holo., BM!, COI! (fragment), G!, L!, K!, P!, iso.)

Stems spreading then pendent to 1–3 m. long, glabrous, iron-grey, densely lenticellate. Leaves mostly opposite; petiole 5–12 mm. long; lamina coriaceous, dull green, linear-lanceolate to elliptic-lanceolate, 5–21 cm. long, 0.7–5 cm. wide, acute to obtuse at the apex, generally cuneate or attenuate at the base, sometimes glandular-hairy at earliest stage of unfolding, with 4–6 pairs of lateral nerves, several lower ones strongly ascending. Umbels several per axil and at older nodes, 2–4-flowered; peduncle 1–5 mm. long; pedicels in sockets, 0–2 mm. long; bract cupular, with small triangular limb, 2–3 mm. long. Receptacle obconic, 1.5–2 mm. long; calyx cupular, 2–4 mm. long, truncate or erose, split by expanding corolla-base. Corolla-tube 3–4 cm. long, pink to crimson outside, generally spotted white, redder along filament-lines inside, head of bud greenish white, darkening, lobes green to black inside; bud-heads oblong-ellipsoid, rounded to truncate, 5-angular, slightly ribbed, lobe-tips sometimes protruding as bosses or points up to 1 mm. long; basal swelling ellipsoid-obovoid, 4–10 mm. long, 3–5 mm. in diameter, the tube constricted for 3–5 mm. above; lobes 9–10 mm. long, reflexed, the upper expanded part 4–5 mm. long, 1–1.5 mm. wide. Stamens red, tooth 0.5–0.7 mm. long, anthers 2.5–3 mm. long. Style with a neck 2.5–3 mm. long; stigma capitate, 0.7–0.8 mm. across. Berry red, oblong-ellipsoid, with persistent calyx, 10–12 mm. long, 6–8 mm. in diameter. Fig. 11/3–8 (p. 76).

TANZANIA. Mpanda District: Kungwe Mt., Kahoko, 23 July 1959, *Newbould & Harley* 4584!; Mbeya District: between Tunduma and Laela, 24 June 1986, *J. & J. Lovett* 847!; Tunduru District: just E. of Songea District boundary, 5 June 1956, *Milne-Redhead & Taylor* 10624!
DISTR. T ?3 (see note), 4, 5, 7, 8; S. Zaire, Angola, Zambia, Malawi, Mozambique and Zimbabwe
HAB. *Brachystegia* woodland, generally on *Brachystegia* or *Julbernardia*, but with local races developing in transition to mixed woodland (see note); 900–1600 m.

SYN. *Loranthus dependens* Engl. in E.J. 20: 117 (1894); Sprague in F.T.A. 6(1): 346 (1910); Balle in F.C.B. 1: 353 (1948); Milne-Redh. in Hook., Ic. Pl. 35, t. 3477 (1950)
  *Acrostephanus dependens* (Engl.) Tiegh. in Bull. Soc. Bot. Fr. 42: 268 (1895)

NOTE. A variable species reaching the edge of its range and infrequently collected in the southern half of Tanzania. Congdon has noted an interesting form of the species that grows on *Lannea schimperi* along scarps from Mufindi to the Rift Valley between Iringa and Mbeya and is specially adapted to fire (*Congdon* 127, 291 & 500). Many shoots arise from the swollen branches of the host and all ascend, even those arising underneath the host branch, in striking contrast to the long pendent stems of the normal form on *Brachystegia*. The leaves are grey-green rather than full green, wavy and brittle, succulent and easily detached. Flowering starts when the shoots are still quite short. *B.D. Burtt* 4792 from the Tubugwe valley in the Mpwapwa District seems to be the same form. On the field label Burtt says he saw it also at Kiberashi in W. Handeni. On *Brachystegia* the leaves can vary considerably from ovate-elliptic to long and strap-shaped.

4. **T. forbesii** (*Sprague*) *Wiens* in Bothalia 12: 423 (1978); Wiens & Tölken in F.S.A. 10: 5, fig. 1/2 (1979); Visser, S. Afr. Parasitic Fl. Pl.: 143, t. 4–5 (1981); Polh. & Wiens, Mistletoes Afr.: 194, photo. 80, fig. 18B (1998). Lectotype, chosen by Wiens (1978): Mozambique, Maputo, Delagoa Bay, *Forbes* (K!, lecto.)

Stems spreading to 1–1.5 m. long, essentially glabrous; lateral branches mostly rather short. Leaves mostly opposite; petiole 4–10 mm. long; lamina coriaceous, green to slightly glaucous, mostly ovate-elliptic to obovate-elliptic, less often lanceolate, 2–8 cm. long, (0.5–)1–4 cm. wide, generally obtuse to shortly rounded at the apex, cuneate to shortly rounded at the base, with 4–6 pairs of lateral nerves, the lower ones strongly ascending. Umbels in axils and at older nodes, 2–4-flowered; peduncle 1–3 mm. long; pedicels 0–3 mm. long; bract saucer-shaped, with a small triangular limb, 1.5–2 mm. long, occasionally slightly glandular-papillate. Receptacle obconic, 1–1.5 mm. long; calyx saucer-shaped, 1–1.5 mm. long. Corolla-tube 2.7–3.2 cm. long, dull red, generally with whitish spots, head of buds green turning greyish black particularly along angles; bud-heads oblong to oblong-obovoid, rounded to subtruncate, 4 mm. long, 2–3 mm. in diameter, angled or narrowly winged; basal swelling variously ellipsoid-obovoid to obovoid or depressed globose, 4–5 mm. long, 3–5 mm. in diameter, the tube constricted for 2–4 mm. above; lobes 7–9 mm. long, reflexed, the upper expanded part 4 mm. long, 1.5–2 mm. wide. Stamens green to red; tooth 0.5–0.7 mm. long; anthers 2–3 mm. long. Style with a neck 2–3 mm. long; stigma capitate, 0.7–0.8 mm. across. Berry not seen, said to be orange.

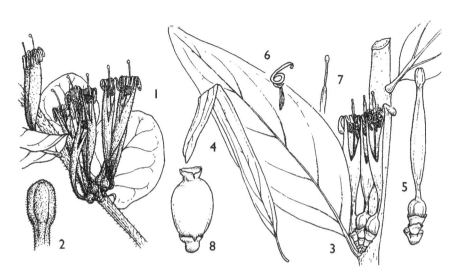

FIG. 11. *TAPINANTHUS ERIANTHUS* — **1**, flowering branch, × 1; **2**, tip of flower bud, × 4. *TAPINANTHUS DEPENDENS* — **3**, flowering node, × 1; **4**, leaf, showing variation, × 1; **5**, flower bud, × 1; **6**, stamen, × 2; **7**, style-tip, × 2; **8**, fruit, × 2. 1, 2, from *Feehan* 2; 3, 5–7, from *Feehan* 88; 4, from *Mendes* 1843. Drawn by Christine Grey-Wilson.

TANZANIA. Lindi District: Lake Lutamba, 12 May 1935, *Schlieben* 6517!
DISTR. **T** 8; W. and S. Angola, SW. Zambia, Malawi, Mozambique, Zimbabwe and north-easternmost South Africa
HAB. Deciduous bushland or thicket; ± 200 m.

SYN. *Loranthus oleifolius* (J.C. Wendl.) Cham. & Schltdl. var. *forbesii* Sprague in Fl. Cap. 5(2): 118 (1915)

NOTE. Closely related to *Tapinanthus dependens*, but easily distinguished by the short calyx. There is a form that occurs on *Uapaca* in a belt from Malawi to SE. Zaire and may be found in SE. Tanzania. It tends to have unusually short peduncles and broad leaves.

5. **T. erianthus** (*Sprague*) *Danser* in Verh. K. Akad. Wet., sect. 2, 29(6): 111 (1933); Polh. & Wiens, Mistletoes Afr.: 196, fig. 17D. Type: Zambia, Central Province, Mine Path near Walamba, *F.A. Rogers* 8353 (K!, holo., B!, iso.)

Stems spreading to 1 m. long; twigs velvety puberulent, sometimes with longer spreading hairs. Leaves opposite, subsessile; petiole 1(–3) mm. long; lamina coriaceous, grey-green to glaucous, ovate, 2–9 cm. long, 2–6 cm. wide, obtuse or rounded at the apex, cordate to amplexicaul at the base, glabrous, with 3–9 pairs of spreading nerves. Umbels 4-flowered; peduncle 1–3 mm. long; pedicels 0–2 mm. long; bract cupular, with a small triangular limb, 2.5–3 mm. long. Receptacle 1–1.5 mm. long; calyx cupular, flared, 1.5–2 mm. long, densely hairy. Corolla-tube 3–4 cm. long, dull to dark red, the lobes green, the heads of buds usually darkening before opening, velvety-puberulous to pilose; bud-heads oblong-obovoid to oblong-ellipsoid, rounded and shortly apiculate to a somewhat truncate apex, slightly angular, 4–5 mm. long, 2.5–3 mm. in diameter; basal swelling ellipsoid-globose to globose, 4–5 mm. long, 3.5–4 mm. in diameter; lobes ± 7 mm. long, reflexed, the upper expanded part oblanceolate, 4–4.5 mm. long, 1.5 mm. wide. Stamens with tooth 0.5–0.7 mm. long; anthers 2–2.5 mm. long. Style with a neck 2–2.5 mm. long; stigma capitate, 0.8–1 mm. across. Berry red, ellipsoid, 8 mm. long, 7 mm. in diameter, with persistent calyx, pubescent. Fig. 11/1, 2.

TANZANIA. Ufipa District: Rukwa to Sumbawanga, ± 6.5 km. from top of Muse Gap, 22 Oct. 1965, *Richards* 20558! & Mbizi Forest Reserve, Nov. 1987, *Ruffo & Kisena* 2792!
DISTR. **T** 4; along Congo-Zambezi divide in S. Zaire (Shaba), Angola, Zambia, Malawi and N. Zimbabwe
HAB. Upland and riverine forest, extending marginally into *Brachystegia* woodland, on a wide variety of hosts; 1650–2150 m.

SYN. *Loranthus erianthus* Sprague in F.T.A. 6(1): 359 (1910) & in K.B. 1911: 149 (1911); Balle in F.C.B. 1: 346, t. A–B facing 346 (1948)

6. **T. globiferus** (*A. Rich.*) *Tiegh.* in Bull. Soc. Bot. Fr. 42: 267 (1895); Danser in Verh. K. Akad. Wet., sect. 2, 29(6): 112 (1933), as '*globifer*'; Balle in F.W.T.A., ed. 2, 1: 662 (1958) & in Fl. Cameroun 23: 54, t. 13/9–10 (1982), pro parte; M.G. Gilbert in Fl. Ethiopia 3: 373, fig. 114.14/3–6 (1990); Polh. & Wiens, Mistletoes Afr.: 201, photo. 12, 84 (1998). Type: Ethiopia, Tigray, near Adowa, Mai Gouagoua, *Quartin-Dillon* (P!, holo.)

Plants with stems spreading to 1 m. or so, glabrous except for the ciliolate bracts and calyx. Leaves mostly opposite or subopposite; petiole 3–10 mm. long; lamina ± coriaceous, green to glaucous, linear-lanceolate to oblong-elliptic, ovate or somewhat oblanceolate, 4–8(–17) cm. long, 0.5–4(–12) cm. wide, acute to obtuse at the apex, cuneate to rounded or shortly cordate at the base, with 4–6 pairs of curved-ascending lateral nerves, the lower ones occasionally more strongly ascending. Umbels mostly in axils, occasionally predominantly at older nodes, (4–)6–8-flowered; peduncle 2–4(–8) mm. long; pedicels 1–3 mm. long from shallow sockets; bract shallowly saucer-shaped, with a small triangular limb, 1.5–2 mm. long. Receptacle barrel-shaped, longer than calyx, 1.5–2 mm. long; calyx shallow, 0.5–1 mm. long. Corolla-tube (2–)2.5–3.5(–4) cm. long, pink to red, rarely white apart from filament-lines, with pale heads generally darkening just prior to anthesis; bud-heads oblong-ellipsoid to narrowly ovoid, angled, ribbed to narrowly winged (wings less than 0.5 mm. wide), occasionally produced into small points at the apex, 3–3.5 mm. long, 1.8–2.5 mm. in diameter; basal swelling ellipsoid-obovoid, 3.5–6 mm. long, 3–4 mm. in diameter, with the tube constricted for 3–5 mm. above; lobes 6–7 mm. long, reflexed, linear-elliptic to lanceolate, 3–4 mm. long, 1–1.2 mm. wide. Stamens white to red; tooth 0.5–0.7 mm. long; anthers 2–2.5 mm. long. Style with neck 2–2.5 mm. long; stigma capitate, 0.8–1 mm. across. Berry orange or red when ripe, ovoid-globose to depressed-globose, 8–9 mm. long, 7–9 mm. in diameter, with a small persistent calyx.

UGANDA. Acholi District: Madi, Aswa [Asua] R., 18 Aug. 1907, *Bagshawe* 1619!
KENYA. Turkana District: Oropoi, Feb. 1965, *Newbould* 7000!
DISTR. **U** 1; **K** 2; widespread from Mauritania to Ethiopia, Djibouti and Yemen
HAB. Wooded grassland; ± 800–1700 m.

SYN. *Loranthus globiferus* A. Rich., Tent. Fl. Abyss. 1: 341 (1848); Sprague in F.T.A. 6(1): 352 (1910); Balle in F.C.B. 1: 346 (1948), pro parte; F.P.S. 2: 293, fig. 105 (1952); E.P.A.: 28 (1953)

## 10. GLOBIMETULA

Tiegh. in Bull. Soc. Bot. Fr. 42: 264 (1895); Polh. & Wiens, Mistletoes Afr.: 209 (1998)

*Loranthus* L. sect. *Symphyanthus* DC. subsect. *Cupulati* DC., Prodr. 4: 298 (1830)
*L.* sect. *Cupulati* (DC.) Sprague in F.T.A. 6(1): 262 (1910)

Stems mostly 0.5–2(–4) m. long, generally from a single attachment, but sometimes scandent with numerous secondary haustoria, hairless but sometimes with papillae on youngest parts and umbels. Leaves opposite, subopposite or ternate, penninerved but second–fourth(–fifth) main lateral nerves often closer, stronger and more ascending. Flowers 2–20 in umbels (often with 1–2 aborted flowers on the peduncle), 5-merous; bracts at the base of peduncle and pedicels small, triangular; flower bract obliquely cupular, with a short triangular or bifid limb, sometimes gibbous or spurred. Calyx cupular or flared, subentire, 0.3–2 mm. long, normally ciliolate. Corolla pink or red, sometimes white to green on upper or lower part, the apical swelling often darkening as bud ripens, 2–4.5 cm. long, with pronounced apical (sometimes winged) and basal swellings, splitting unilaterally more than halfway, the lobes coiled outwards. Stamens involute; anthers emarginate, with a distinct connective, and the outer pair of thecae shorter and sometimes set lower. Style skittle-shaped above; stigma turbinate to peltate. Berry red to yellow, depressed-globose to ellipsoid; seed also brightly coloured, similar to *Tapinanthus*.

13 species in tropical Africa.

Differs from *Tapinanthus* by its pollination mechanism; the apical swelling of the flower-bud signals maturity by changing colour and when prodded the petals curl back to expose the fertile column; further probing of a secondary vent at the base of the filaments causes the corolla to rip and the stamens to flick inwards scattering the pollen explosively. Its closest affinity may be with *Moquiniella* Balle from the Cape region of southern Africa.

1. Corolla-buds red or pink at base only, grey-white above (head darkening at maturity), with obovoid ± truncate apical swelling (fig. 12/9); leaves glaucous · · · · · · · · · ·    4. *G. rubripes*
   Corolla mainly red, head sometimes green; apical swelling of heads globular to subhemispherical; leaves not glaucous · · · · · · · · · · · · 2
2. Leaves pointed to acuminate, with midvein penetrating to tip, 2.5–4 times as long as broad · · · · · · · · · · · · · · · ·    5. *G. kivuensis*
   Leaves bluntly pointed to rounded, with midvein divaricating into subequal branches well before tip, 1.5–2.5 times as long as broad · · · · · · · · · · · · · · · · · · · · · · · · · · · · · 3
3. Heads of corolla-buds 5–7 mm. across, ± winged; stigma peltate, 3 mm. across · · · · · · · · · · · · · · · · · · · · · · · ·    3. *G. pachyclada*
   Heads of corolla-buds 2.5–3.5 mm. across, the lobe-margins not or only slightly reduplicate; stigma 1–2 mm. across · · · · · · · · · · · · · 4
4. Calyx 0.5–0.7 mm. long; filaments and inside of corolla-lobes red to purplish; bark of twigs not breaking down · · · · · · ·    1. *G. braunii*
   Calyx (1–)1.5–2 mm. long; filaments and inside of corolla-lobes green; twigs becoming powdery on previous season's growth · · · · · · · · · · · · · · · · · · · · · · · · · · · · · · · ·    2. *G. anguliflora*

1. **G. braunii** (*Engl.*) *Danser* in Verh. K. Akad. Wet., sect. 2, 29(6): 54 (1933); Balle in F.W.T.A., ed. 2, 1: 660 (1958), excl. specim. *Kitson*; Balle & Hallé in Adansonia, sér. 2, 1: 227 (1962); Balle in Fl. Cameroun 23: 15, fig. 3/1–10 (1982); Polh. & Wiens in U.K.W.F., ed. 2: 157, t. 56 (1994) & Mistletoes Afr.: 211, photo. 88, fig. 21A. Type: Equatorial Guinea, Rio Muni, *Mann* 1833* (B!, holo., K!, iso.)

Twigs terete or elliptic in section, glabrous. Leaves opposite, subopposite or sometimes (in Zaire basin) ternate; petiole 5–15 mm. long; lamina fleshy, reddish when young, turning light to dark green, elliptic, elliptic-oblong or elliptic-ovate, 4–10 cm. long, 1.5–7 cm. wide, blunt or rounded at the apex, broadly cuneate to shortly cordate at the base; second and third main lateral nerves curved-ascending to upper half, the distal ones more spreading. Umbels 3–6, mostly at older leafless nodes; peduncle 3–20 mm. long, glabrous or sometimes scurfy; pedicels 4–9(–15), 2.5–5 mm. long; bract-cup 0.5 mm. long, limb triangular or bifid, 0.5–1.2 mm. long.

* Erroneously quoted as 1883 by Engler in the protologue.

Receptacle 1–1.5 mm. long; calyx 0.5–0.7 mm. long. Corolla red or pink, the apical swelling darkening in ripe buds, 2.5–3.5 cm. long; apical swelling of mature buds subhemispherical, slightly pointed, pentagonal, slightly reduplicate proximally, 2.5–3.5 mm. across, 2–2.5 mm. tall; basal swelling usually subglobose, 2–4 mm. long, 2–3 mm. in diameter, rarely ellipsoid. Inner anthers 1.5–2 mm. long; outer ones 1–1.5 mm. Style-constriction 1.5–2 mm. long; stigma peltate to depressed-turbinate, 1.5–2 mm. across. Berry orange or red, 5–6 mm. long, 5 mm. in diameter; seed orange or red, 3 mm. long, 1.5 mm. in diameter. Fig. 12/1–8 (p. 81).

UGANDA. Masaka District: 1 km. SE. of Nkoma, 13 June 1971, *Lye* 6241!; Mengo District: near Entebbe, Kitabi, Oct. 1935, *Chandler* 1941 & near Buikwo, 21 Aug. 1949, *Ocrmarton* 3091

KENYA. N. Kavirondo District: Webuye [Broderick] Falls, Mar. 1961, *Tweedie* 2110! & Malava, near Mlaba Forest, 7 Oct. 1982, *R.M. & D. Polhill* 4861! & Kakamega Forest, 27 Aug. 1975, *Kokwaro* 3780!

TANZANIA. Bukoba District: Minziro, 22 Nov. 1994, *Congdon* 386!; Buha District: Gombe Stream Reserve, Kakombe valley, 10 Jan. 1964, *Pirozynski* 222!

DISTR. U 4; K 5; T 1, 4; Guineo-Congolian forests from Ivory Coast to S. Sudan

HAB. Evergreen and riverine forest, often at edges or in cleared places nearby, on a wide variety of hosts and prone to invade plantation crops; 1050–1600 m.

SYN. *Loranthus braunii* Engl. in E.J. 20: 93 (1894); Sprague in F.T.A. 6(1): 303 (1910) & 1028 (1913); Balle in F.C.B. 1: 369 (1948), excl. syn. *L. mweroensis*; U.K.W.F.: 332, fig. on 333 (1974)

    *L. sp.* sensu F.P.U.: 100, t. 15 (1962)

NOTE. This is the most widespread species of the genus, more variable than the others and not always easy to distinguish from some of them.

2. **G. anguliflora** (*Engl.*) *Danser* in Verh. K. Akad. Wet., sect. 2, 29(6): 54 (1933); Polh. & Wiens, Mistletoes Afr.: 213, photo. 89A–B (1998). Type: Angola, Huila, Lake Ivantala, *Welwitsch* 4887 (B!, holo., BM!, COI!, G!, K!, LISU!, iso.)

Twigs terete or elliptic in section, glabrous; bark smooth, grey, becoming powdery. Leaves opposite or subopposite; petiole 5–15(–20) mm. long; lamina leathery or slightly fleshy, dull green, slightly paler beneath, often tinged reddish when young, elliptic to ovate, 4–14 cm. long, 2–6.5 cm. wide, bluntly pointed at the apex, cuneate at the base, slightly decurrent; main lateral nerves 3–6(–7) pairs, second to fourth (or fifth) rather strongly ascending, usually well spaced but sometimes arising nearly together at the base. Umbels 3–15, mostly at nodes below current leaves; peduncle 2–15 mm. long, glabrous or shortly hairy; pedicels 3–6, 2–4 mm. long; bract-cup 0.5–1.5 mm. long, with a triangular or bifid limb 0.5–3 mm. long. Receptacle 1–1.5 mm. long; calyx (1–)1.5–2 mm. long. Corolla red, the apical swelling usually green and darkening in ripe buds, but lobes green inside, 3–4 cm. long; apical swelling of mature buds subglobose to subhemispherical, not or slightly reduplicate, pentagonal, 2.5–3 mm. across, 2.5 mm. tall; basal swelling ellipsoid, 3–4 mm. long, 2–3 mm. in diameter. Inner anthers 1–2 mm. long; outer ones 0.5–1.5 mm. Style-constriction 1–2 mm. long; stigma peltate to depressed-turbinate, 1–1.8 mm. across. Berry ripening yellow, ellipsoid to obovoid-truncate, 9–10 mm. long, 6–9 mm. in diameter.

TANZANIA. Kigoma District: Usondo Plateau, 9 Feb. 1997, *Congdon* 489!; Njombe District: Njombe–Songea road, Madaba, 27 Jan. 1985, *Congdon* 341!; Songea District: 5 km. E. of Songea, Unangwa Hill, 15 Jan. 1956, *Milne-Redhead & Taylor* 8335!

DISTR. T 4, 7, 8; Zaire (Shaba), Angola, Zambia, Malawi and Mozambique

HAB. Higher rainfall deciduous woodland, commonly on *Uapaca*, but also on *Brachystegia* and other mostly leguminous trees; 900–1200 m.

SYN. *Loranthus anguliflorus* Engl. in E.J. 20: 107 (1894); Sprague in F.T.A. 6(1): 304 (1910)

NOTE. Generally easily distinguished from *G. braunii* by the longer calyx, green-headed flower-buds (corolla-lobes greenish inside) and powdery bark on the older branches.

3. **G. pachyclada** (*Sprague*) *Danser* in Verh. K. Akad. Wet., sect. 2, 29(6): 55 (1933); Polh. & Wiens, Mistletoes Afr.: 214, photo. 91, fig. 21E (1998). Type: Tanzania, Njombe District, Ukinga, Mt. Tyuni, *Goetze* 1003 (B!, holo., BR!, P!, iso., K!, fragment)

Twigs terete or elliptical in section, glabrous; bark grey-brown, not peeling. Leaves opposite or subopposite; petiole 5–8 mm. long; lamina thick and fleshy, yellow-green, tinged reddish when young, elliptic to broadly ovate, 3.5–9 cm. long, 2–5.5 cm. wide, bluntly pointed to rounded at the apex, obliquely cuneate to rounded and slightly decurrent at the base; main lateral nerves 6–8 pairs, second to fourth strongly ascending, the upper ones more spreading. Umbels 3–6, mostly at older leafless nodes; peduncle 3–8 mm. long; pedicels 3–6, 3–6 mm. long; bract-cup 0.5 mm. long, with a triangular or bifid limb 1–3 mm. long. Receptacle 0.7–1.2 mm. long; calyx 1.2–2 mm. long. Corolla bright deep pink, the apical swelling turning black in ripe buds, the lobes green inside, 3.5–4.5 cm. long; apical swelling subhemispherical, pentagonal, reduplicate to narrowly 5-winged, the wings extending down the tube, 5–7 mm. across, 4–5 mm. tall; basal swelling slight. Inner anther-thecae 5 mm. long; outer ones 4 mm. long. Style-constriction 5 mm. long; stigma peltate, 3 mm. across. Berry mauvish, turbinate, to 12 mm. across. Fig. 12/10.

TANZANIA. Mbeya District: Pungaluma Hills, above Mshewe village, 21 Mar. 1990, *Lovett & Kayombo* 4385!; Iringa District: 22 km. N. of Mafinga [Sao Hill] turnoff, 13 July 1956, *Milne-Redhead & Taylor* 11064!; Njombe District: Chimala–Matamba road, above Ndumbi gorge on way to Kitulo Plateau, 25 May 1986, *Congdon* 86!
DISTR. **T** 7; not known elsewhere
HAB. *Brachystegia-Uapaca* woodland, on *Uapaca*; 1300–2200 m.

SYN. [*Loranthus anguliflorus* sensu Engl. in E.J. 30: 301 (1901); F.D.O.-A. 2: 163 (1932), *non* Engl. (1894)]
   *L. pachycladus* Sprague in K.B. 1912: 233 (1912) & in F.T.A. 6(1): 1028 (1913); F.D.O.-A. 2: 178 (1932); T.T.C.L.: 286 (1949)

NOTE. Easily recognised by the exceptionally large flowers and fruits.

4. **G. rubripes** (*Engl. & K. Krause*) *Danser* in Verh. K. Akad. Wet., sect. 2, 29(6): 55 (1933); Polh. & Wiens, Mistletoes Afr.: 215, photo. 92, fig. 21D (1998). Types: Tanzania, Rungwe District, Bulambia, Mt. Milambi, *Stolz* 1614 (B†, syn., BM!, BR!, K!, isosyn.) & Songwe R., *Stolz* 1616 (B†, syn., BM!, EA!, FHO!, K!, MO!, isosyn.)

Twigs terete or elliptic in section, glabrous or scurfy on youngest parts; bark grey, rarely powdery. Leaves opposite or subopposite; petiole 5–30 mm. long; lamina leathery or slightly fleshy, glaucous, ± red flushed when young, elliptic-oblong, elliptic or ovate, 4–10 cm. long, 2–7.5 cm. wide, bluntly pointed at the apex, narrowly to broadly cuneate and slightly decurrent at the base; main lateral nerves 4–7 pairs, second to fourth strongly ascending, the well-separated upper ones more spreading, or in broad leaves all more regularly spreading. Umbels 1–3 in axils and at older leafless nodes; peduncle (2–)5–35 mm. long, sometimes with a few aborted buds below umbel; pedicels 3–9(–18), (2–)3–5 mm. long, glabrous or hairy; bract-cup 0.5–1 mm. long, with a triangular or bifid limb 0.5–1.5 mm. long. Receptacle 1–1.2 mm. long; calyx 0.3–0.7 mm. long. Corolla red or pink at the base, white or grey above, the apical swelling darkening in ripe buds, lobes green or whitish inside, 2.5–4.5 cm. long; apical swelling obovoid, ± truncate, pentagonal, slightly reduplicate, 2.5–3 mm. across; basal swelling ellipsoid, (3–)4–6 mm. long, (2.5–)3.5–4 mm. in diameter. Inner anther-thecae 2–4 mm. long; outer ones 0.5–2.5 mm. Style-constriction 2–4 mm. long; stigma peltate 1.2–1.5 mm. across. Berry obovoid-globose, 10–12 mm. long, 9–10 mm. in diameter, glaucous turning dull purple, slightly verruculose. Fig. 12/9.

TANZANIA. Mbulu District: 95 km. N. of Kondoa on road to Arusha, 13 Oct. 1988, *Wiens & Calvin* 7018!; Njombe Disrict: Chimala–Matamba road on escarpment, 18 Nov. 1989, *Congdon* 227!; Songea District: 16 km. S. of Gumbiro, 27 Jan. 1956, *Milne-Redhead & Taylor* 8562!

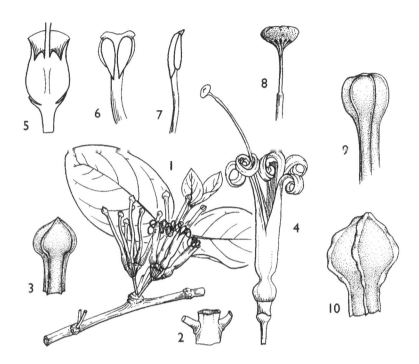

FIG. 12. *GLOBIMETULA BRAUNII* — **1**, flowering branch, × ²/₃; **2**, detail of node × 2; **3**, tip of flower bud, × 4; **4**, flower, × 2; **5**, section of base of flower, × 6; **6, 7**, anthers, front and side views, × 6; **8**, style-tip, × 6. *GLOBIMETULA RUBRIPES* — **9**, tip of flower bud, × 4. *GLOBIMETULA PACHYCLADA* — **10**, tip of flower bud, × 4. 1, 2, from *Tweedie* 1544; 3–8, from *Tweedie* 1929; 9, from *Milne-Redhead & Taylor* 8562; 10, from *Milne-Redhead & Taylor* 11064. Drawn by Christine Grey-Wilson

DISTR. **T** 2, 5, 7, 8; Angola, Zambia, Malawi and Mozambique

HAB. Higher rainfall deciduous woodland, generally on *Brachystegia* or other legumes, also on *Parinari*; 900–1650 m.

SYN. *Loranthus rubripes* Engl. & K. Krause in E.J. 51: 458 (1914); F.D.O.-A. 2: 164 (1932); T.T.C.L.: 289 (1949)

5. **G. kivuensis** (*Balle*) *Polh. & Wiens* in Lebrun & Stork, Énum. Pl. Fl. Afr. Trop. 2: 168 (1992); Balle & Troupin in Fl. Rwanda 1: 185, fig. 38/2 (1978), *nom. invalid.*; Troupin, Fl. Pl. Lign. Rwanda: 374, fig. 132/2 (1982), *nom. invalid.*; Polh. & Wiens, Mistletoes Afr.: 216 (1998). Type: Zaire, Kivu, Beni–Lubero, *Lebrun* 4293 (BR!, holo.)

Twigs compressed or 4-ribbed. Leaves opposite, subopposite or ternate, purplish when young; petiole 6–15 mm. long, sometimes reddish; lamina thinly coriaceous, oblong elliptic, 6–16 cm. long, 2–3.5 cm. wide, pointed to acuminate at the apex, cuneate and slightly decurrent at the base, with 7–10 pairs of well-spaced curved-ascending lateral nerves. Umbels 2–6, mostly at nodes below current leaves; peduncle 2–13 mm. long; pedicels 4–8, 2–4 mm. long; bract-cup 0.7–1 mm. long, slightly gibbous, with a triangular or bifid limb 1–2 mm. long, ciliolate. Receptacle 1.2–1.5 mm. long; calyx 1–1.5 mm. long. Corolla pink, sometimes paler at extremities, the apical swelling turning blue-black before opening, 2.7–3.5 cm. long; apical swelling subhemispherical, slightly pointed, pentagonal, slightly reduplicate proximally, 3 mm. across, 2 mm. tall; basal swelling subglobose, 2–3 mm. long. Inner anther-thecae 1.5–2 mm. long; outer ones 1–1.2 mm. long. Style constriction 1.5–2 mm. long; stigma peltate, 1.2–1.5 mm. across. Berry globose or ellipsoid, ± 6 mm. long.

UGANDA. Ankole District: Kalinzu Forest, June 1938, *Eggeling* 3706!; Kigezi District: Kayonza Forest, 1 June 1957, *V.G. van Someren* 2!
DISTR. **U** 2; E. Zaire and Rwanda
HAB. Upland evergreen forest margins, on various hosts; 1350–1600 m.

SYN. *Loranthus mweroensis* (Baker) Danser var. *kivuensis* Balle in B.J.B.B. 17: 232 (1944)
    *L. kivuensis* (Balle) Balle in F.C.B. 1: 373 (1948); F.P.N.A. 1: 101 (1948)

NOTE. The flowers are similar to those of *G. braunii* apart from a rather longer calyx, but the narrow pointed thinly textured leaves provide an easy distinction.

## 11. TAXILLUS

Tiegh. in Bull. Soc. Bot. Fr. 42: 256 (1895); Polh. & Wiens, Mistletoes Afr.: 219 (1998)

Shrubs, rarely exceeding 1 m. in height, with haustoria bearing surface runners; hairs stellate and dendritic, sometimes with other simple ones; twigs terete to angular. Leaves alternate to opposite, petiolate; lateral nerves pinnate, principally 3–5 from the base, or obscure. Flowers axillary or terminal on short shoots in umbels or clusters, sometimes sessile, occasionally solitary; bract single, unilateral or cupular, entire or toothed. Corolla 4–5-merous, often hairy, opening spontaneously or usually explosively, zygomorphic, with a V-split on one side, generally yellow or red and green; buds often curved, somewhat inflated medially, often narrowed below the slightly ellipsoid to clavate head, sometimes developing vents below the head, often with a gland on one side; lobes shorter than tube, erect (in Africa) or reflexed, linear to spathulate. Stamens attached near the base of the corolla-lobes, erect; filaments short, straight; anthers 4-thecous, sometimes locellate, with a very small connective-appendage. Style slender, terete to angular; stigma small. Berry usually reddish, ovoid, ellipsoid or obovoid, smooth or verruculose, with a persistent calyx.

About 35 species in tropical Asia, from India and Sri Lanka to China, Japan, Philippines and Borneo, with one species on the Kenya coast.

*Taxillus wiensii* was discovered only in 1972 and despite several visits to the type locality in the Sokoke Forest of coastal Kenya it has never been found flowering more than sporadically. The general appearance is certainly very similar to species of *Taxillus* seen live in Sri Lanka (Wiens in Dassan., Rev. Handb. Fl. Ceylon 6: 138 (1988)), but the flowers seem less specialised than those of the Asiatic species. The flowers may open spontaneously or by some manipulation of the head in bud by birds. They open with a short V-slit, the style springing forward, the filaments short and erect.

**T. wiensii** *Polh.* in Polh. & Wiens, Mistletoes Afr.: 219, photo. 94, fig. 22 (1998). Type: Kenya, Kilifi District, 20 km. Malindi–Jilore, *Wiens* 4526 (K!, holo., BR!, EA!, K!, LISC!, SRGH!, iso.)

Stems to 50 cm. from an extensive untidy system of surface runners bearing numerous haustorial connections; twigs terete, tomentellous with brownish dendritic hairs, soon glabrescent. Leaves opposite to subopposite; petiole 1–2 mm. long; lamina coriaceous, mid-green, elliptic-obovate, 2–3.5 cm. long, 1–2 cm. wide, rounded at the apex, cuneate to rounded at the base, tomentose on both surfaces, soon glabrescent, obscurely nerved. Flowers 1–6, sessile, clustered in the axils, 5-merous; bract unilateral, narrowly boat-shaped, 2–3 mm. long, glabrescent. Receptacle urceolate, 1.5–2 mm. long, tomentose; calyx annular, 1 mm. long, long-ciliate. Corolla 3 cm. long, dull purplish, lobes green turning dull red inside, white

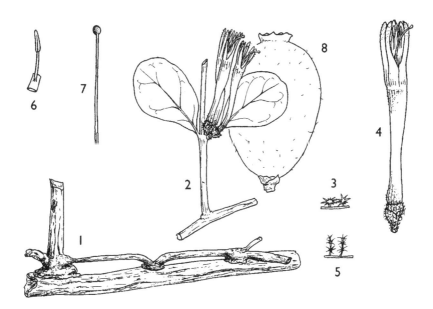

FIG. 13 *TAXILLUS WIENSII* — **1**, haustoria and surface runners, × ²/₃; **2**, flowering branch, × 1; **3**, hairs from twig, × 20; **4**, flower, × 2; **5**, hairs from flower, × 20; **6**, stamen, × 4; **7**, style-tip, × 4; **8**, fruit, × 2. 1, 3, 8, from *Polhill* 4825; 2, 4–7, from *Wiens* 4526. Drawn by Christine Grey-Wilson.

stellate-pubescent; buds with a narrow straight tube, with a slight ellipsoid-clavate apical swelling; lobes linear, slightly thickened, 4–5 mm. long, remaining erect. Anthers 1–1.5 mm. long, as long as the filaments. Style dull purple, slender, tapered; stigma ovoid-subglobose, 0.6 mm. across. Berry red, ovoid, 7 mm. long, 5 mm. in diameter. Fig. 13.

KENYA. Kilifi District: 20 km. on Malindi–Jilore road, 24 Sept. 1982, *Polhill* 4825! & SW. corner of Arabuko-Sokoke Forest, 27 Mar. 1990, *Luke & S.A. Robertson* 2173! & 16 Nov. 1993, *S.A. Robertson & Polhill* 6861!
DISTR. **K** 7; known only from the Arabuko-Sokoke Forest
HAB. Dry evergreen coastal forest, on *Cynometra webberi*; 100–150 m.

NOTE. As indicated above more detailed studies of the pollination biology of this relict population would be valuable.

## 12. VANWYKIA

Wiens in Bothalia 12: 422 (1978); Polh. & Wiens, Mistletoes Afr.: 221 (1998)

*Loranthus* L. sect. *Remoti* Sprague in F.T.A. 6(1): 265 (1910)

Medium-sized to large shrubs to 1.5 m. or so, spreading by stout haustoria-bearing surface runners; hairs stellate and dendritic; twigs terete. Leaves alternate to opposite, sometimes fascicled, petiolate, penninerved. Flowers 2–6 in sessile to shortly stalked umbels in the axils and crowded at older nodes; bract unilateral, lanceolate to ovate. Calyx annular. Corolla (4–)5-merous, green turning reddish, tomentose, bilaterally symmetrical, developing only a short V-slit; buds straight, with ellipsoid to obovoid head, induplicate; lobes short, erect, somewhat incurved at the

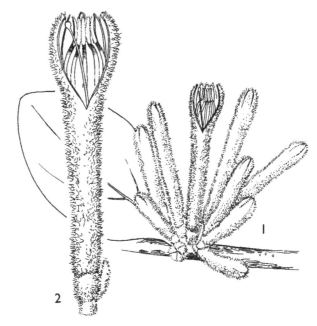

FIG. 14. *VANWYKIA REMOTA* — **1**, flowering node, × 1; **2**, flower, × 2. Drawn by Marguerite Scott. Reproduced from Flora of Souhern Africa.

tip, spathulate. Stamens attached at the base of corolla-lobes, short, essentially erect, but curving inward to form a central, collective anther-mass; anthers 4-thecous, with minute connective appendage. Style terete, tapered, pubescent on lower half; stigma obovoid to globular. Berry red, ellipsoid to cylindrical, hairy.

Two species in eastern and south-eastern Africa.

Closely related to *Taxillus* and included there by Balle in Webbia 11: 580 (1955). The flowers of *Vanwykia* are weakly explosive at the most (Feehan in J.L.S. 90: 140 (1985) and Kirkup in Polh. & Wiens, Mistletoes Afr.: 42 (1998)). The swollen bud may be squeezed by sunbirds or the corolla-lobes may separate spontaneously in a V-slit to reveal the anthers grouped in a column around the pistil. The scarlet anther-backs function as a honey-guide encouraging a pollinator to probe further, causing the rent in the corolla to extend further so that the short stamens are weakly inflexed and the freed style tilts forwards. After pollination the red coloration spreads further down the stamens, and the style also reddens.

Hairs white except sometimes for a brownish tinge on bud-heads, 2–4 mm. long on the corolla-tube; leaf-lamina elliptic-ovate to elliptic-obovate or subcircular, 1.2–2.2 times as long as broad · · · · · · · · · · · · · · · · · · · · · · · · · · · · · · · · · · · · · · 1. *V. remota*
Hairs mostly reddish brown to bright purplish red, mostly 1–2 mm. long on the corolla-tube; leaf-lamina linear-lanceolate to oblong-elliptic, 3–8 times as long as broad · · · · · · · · · · · · · · 2. *V. rubella*

1. **V. remota** (*Baker & Sprague*) *Wiens* in Bothalia 12: 422 (1978); Wiens & Tölken in F.S.A. 10: 29, fig. 11/3 (1979); Vollesen in Opera Bot. 59: 64 (1980); Germish. et al. in Fl. Pl. Afr. 53, t. 2081 (1994); Polh. & Wiens, Mistletoes Afr.: 221, photo. 95, fig. 23 (1998). Type: Mozambique, Chupanga [Shupanga], *Kirk* 40 (K!, holo.)

Stems to 1–1.5 m. long, becoming pendent; twigs tomentose with whitish hairs 1–2 mm. long. Leaves opposite to subopposite, occasionally alternate or clustered on short shoots; petiole 5–10 mm. long; lamina coriaceous, elliptic-obovate to elliptic-ovate or subcircular, 3–9 cm. long, 1.5–4 cm. wide, rounded at the apex, cuneate and attenuate at the base, tomentellous at first, soon glabrescent. Umbels in axils and around old nodes, 2–5-flowered, tomentose; peduncle 0–3 mm. long; pedicels 1–3 mm. long; bract 3–4 mm. long. Receptacle 3–4 mm. long, tomentose; calyx scarcely differentiated. Corolla (4–)5-merous, 4–5 cm. long, yellow-green and whitish tomentose outside with hairs 2–4 mm. long, green turning reddish and sometimes pubescent inside; lobes spathulate, 8–10 mm. long. Anthers 2.5–4 mm. long, scarlet. Berry red, cylindrical when mature, 12–19 mm. long, 0–7 mm. in diameter, tomentose. Fig. 14.

TANZANIA. Kondoa District: Narai, Nov. 1925, *B.D. Burtt* 1484!; Morogoro District: 97 km. NW. of Morogoro on new Dodoma road, Kimali Hill, 8 June 1990, *Carter, Abdallah & Newton* 2650A!; Kilwa District: Selous Game Reserve, Njenje, 26 Aug. 1971, *Rees* T160!
DISTR. **T** 3, 5, 6, 8; E. Zambia, Malawi, Mozambique and north-easternmost South Africa
HAB. Deciduous woodland and elsewhere in riverine forest, often on *Ficus*, but also on other hosts; 200–1400 m.

SYN. *Loranthus remotus* Baker & Sprague in F.T.A. 6(1): 327 (1910) & in K.B. 1911: 143 (1911)
    *Tapinanthus remotus* (Baker & Sprague) Danser in Verh. K. Akad. Wet., sect. 2, 29(6): 118 (1933)

2. **V. rubella** *Polh. & Wiens*, Mistletoes Afr.: 222 (1998). Type: Tanzania, Kahama District, Ngomero Mt., *Bullock* 3135 (K!, holo.)

Stems forming clumps to 1 m. or so; twigs tomentellous, longer hairs reddish and up to 1 mm. long. Leaves mostly opposite to subopposite; petiole 5–15 mm. long; lamina coriaceous, linear-lanceolate to narrowly oblong-elliptic, 6–11 cm. long, 1–3.5 cm. wide, bluntly pointed at the apex, cuneate at the base, reddish tomentellous, soon glabrescent. Umbels in axils and at older nodes, 2–5-flowered, tomentellous; peduncle 0–2 mm. long; pedicels 0–2(–3 in fruit) mm. long; bract 2–4 mm. long. Receptacle 3–3.5 mm. long, tomentellous; calyx scarcely differentiated. Corolla 5-merous, 4–4.5 cm. long, cream turning salmon-pink and with reddish-brown to purplish-red hairs up to 2 mm. long outside, green turning red and sometimes hairy inside; lobes oblanceolate-spathulate, 6–8 mm. long. Anthers 3 mm. long. Berry becoming ellipsoid, 13–15 mm. long, 5–6 mm. in diameter, tomentellous.

TANZANIA. Ufipa District: road to Kasanga, 13 June 1957, *Richards* 10091!; Singida District: Rift Escarpment, Sept. 1935, *B.D. Burtt* 5279!; Mbeya District: Madibira, 22 May 1988, *Congdon* 207!
DISTR. **T** 4, 5, 7; N. Zambia
HAB. Deciduous woodland, on *Brachystegia* and other legumes; 1000–1800 m.

SYN. [*Loranthus remotus* sensu T.T.C.L.: 288 (1949), pro parte, *non* Baker & Sprague]

## 13. OEDINA

Tiegh. in Bull. Soc. Bot. Fr. 42: 249 (1895); Polh. & Wiens, Mistletoes Afr.: 225 (1998)

*Loranthus* L. sect. *Dendrophthoë* (Mart.) Engl. group *Laxiflori* Engl. in E.J. 20: 99 (1894)
    L. sect. *Laxiflori* (Engl.) Sprague in F.T.A. 6(1): 262 (1910)
*L.* subgen. *Dendrophthoë* (Mart.) Engl. sect. *Botryoloranthus* Engl. & K. Krause in E.J. 51: 461 (1914)
    *L.* sect. *Botryoloranthus* (Engl. & K. Krause) De Wild., Pl. Bequaert 1: 301 (1922)
[*Dendrophthoë* sensu Danser in Verh. K. Akad. Wet., sect. 2, 29(6): 20, 43 (1933), *non* Mart.]
*Botryoloranthus* (Engl. & K. Krause) Balle in Bull. Séances Acad. Roy. Sci. Col. 25: 1622 (1954) & in Webbia 11: 580 (1955)

Shrubs to 2 m. or more from a single haustorial attachment; hairs stellate and dendritic; twigs slightly compressed to angular, soon terete. Leaves mostly opposite or subopposite, rarely clustered, shortly petiolate, penninerved. Flowers in racemes or spikes from axils and older nodes, few–many-flowered; bract small, unilateral. Calyx saucer-shaped to tubular, entire to shortly toothed. Corolla joined up to halfway, 5-lobed, radially symmetrical to slightly zygomorphic, yellow to red without any marked colour-banding, puberulous to hirsute, weakly explosive at anthesis; buds opening first by vents below the middle, only slightly expanded over the anthers; tube with a slight to marked basal swelling, only rarely with a short V-slit at anthesis; lobes usually erect, often partly cohering at tips, some or all sometimes spreading to slightly reflexed from near point of filament-insertion, occasionally coiled and tending to break off. Filaments attached 6–10 mm. above the base of corolla-lobes, slender, the upper part slightly thickened, ± articulate, the upper part coiling, collapsing and breaking off to various degrees; anthers basifixed, 4-thecous; connective-appendage minute to subulate. Style filiform; stigma ovoid to fusiform. Berry, where known, blue-green, ovoid-ellipsoid to obovoid; seed orange to red.

4 species in Tanzania and Malawi.

*Oedina* is closely allied to Asian genera, particularly *Dendrophthoë*, and is relatively unspecialised among the African genera. The few species have rather restricted distributions in the montane forests from Tanzania to northern Malawi, and are only locally common. The species are quite diverse in a number of features that become stabilised in more advanced genera and the limits between *Oedina*, *Oncella* and *Erianthemum* remain somewhat problematic, as discussed by Polhill & Wiens (1998).

The flowers are opened by probing vents above a relatively short corolla-tube, the lobes in *O. erecta* remaining erect (often somewhat coherent on the side opposite to that opened) to somewhat spreading, in *O. pendens* coiling and mostly breaking off. The weakly explosive opening propels pollen in one direction towards the pollinator and the slightly thicker upper part of the filaments coil, collapse and mostly detach. The species of *Dendrophthoë* in Asia all have relatively short filaments inserted nearer the base of the corolla-lobes and remain erect; the corolla is opened variously at vents or by pinching the tip of the mature bud.

1. Flowers with at least a short pedicel; calyx less than 1 mm. long · · · · · · · · · · 2
   Flowers sessile; calyx tubular, 2–4 mm. long · · · · · · · · · · · · · · · · · · · · · · 3
2. Corolla 5.5–6.5 cm. long, orange below, red above, the lobes coiled and tending to break off at anthesis; petioles 10–30 mm. long · · · · · · · · · · · · · · · · · · · · · · · · ·    1. *O. pendens*
   Corolla 2.5–3 cm. long, yellow with rusty-brown hairs, the lobes erect or spreading to reflexed from the middle; petioles 1–5 mm. long · · · · · · · · · · · · · · · · · · · ·    2. *O. erecta*
3. Corolla orange, stellate pubescent, without a V-slit in tube; leaves soon glabrescent · · · · · · · · · · · · · · · · · · · · · ·    3. *O. brevispicata*
   Corolla dull yellow, hirsute with rusty-brown hairs, developing a short V-slit in tube; leaves rusty tomentellous beneath · · · · · · · · · · · · · · · · · · · · · · · · · · · · · ·    4. *O. congdoniana*

1. **O. pendens** (*Engl. & K. Krause*) *Polh. & Wiens* in Lebrun & Stork, Énum. Pl. Fl. Afr. Trop. 2: 169 (1992) & Mistletoes Afr.: 227, photo. 98, fig. 24A (1998). Type: Tanzania, Rungwe District, Mt. Rungwe, *Stolz* 1103 (B!, holo., K!, WAG!, iso.)

Shrub with spreading tangled branches to 2 m. or more, with fine short dendritic hairs on youngest parts, but soon glabrescent. Leaves opposite or subopposite, occasionally clustered on short shoots; petiole 1–3 cm. long; lamina thinly coriaceous, lanceolate to oblong-elliptic or ovate, 4–20 cm. long, 2–8 cm. wide, bluntly pointed to rounded at the apex, rounded to shortly cordate at the base; lateral nerves 8–10 pairs. Racemes axillary and mostly at older nodes, 2.5–12 cm. long, with numerous flowers; pedicel 0.5–4 mm. long; bract ovate-triangular, concave, 1.5–2 mm. long, ciliate. Receptacle 2–3 mm. long; calyx 0.5–0.8 mm. long, ciliolate. Corolla 5.5–6.5 cm. long, with tube orange at least below, upper part fiery red, puberulous, glabrescent, slightly curved in bud; tube 2–2.5 cm. long, slightly swollen at the base; lobes coiling above the vented region at anthesis and some breaking off, the pollen expelled unilaterally but tube not split. Filaments inserted ± 1 cm. above the base of the lobes, slender, straight at first but collapsing within seconds of anthesis, breaking off a few mm. above broken corolla-lobes; anthers 5–6 mm. long; connective-appendage subulate, 1.5–2 mm. long. Berry blue green, obovoid, 10 mm. long, 0 mm. in diameter, seed dull red. Fig. 15/1–7 (p. 88).

TANZANIA. Mbulu, 15 Apr. 1954, *Matalu* 3078!; Morogoro District: Uluguru Mts., Salaza Forest 5 km. S. of Bunduki, 15 Mar. 1953, *Drummond & Hemsley* 1626!; Iringa District: Mufindi, Lugoda Tea Estate, Kibao, 5 Jan. 1986, *D. & C. Wiens & Congdon* 6544!
DISTR. **T** 2, 6, 7; Zambia and N. Malawi
HAB. Montane forest edges, on various hosts; 1350–2300 m.

SYN. *Loranthus pendens* Engl. & K. Krause in E.J. 51: 461, fig. 2 (1914); F.D.O.-A. 2: 164 (1932); T.T.C.L.: 286 (1949)
  *Dendrophthoë pendens* (Engl. & K. Krause) Danser in Verh. K. Akad. Wet., sect. 2, 29(6): 47 (1933)
  *Botryoloranthus pendens* (Engl. & K. Krause) Balle in Bull. Séances Acad. Roy. Sci. Col. 25: 1624 (1954)

2. **O. erecta** (*Engl.*) *Tiegh.* in Bull. Soc. Bot. Fr. 42: 249 (1895); Polh. & Wiens, Mistletoes Afr.: 227, photo. 99, fig. 24C (1998). Types: Tanzania, Lushoto District, W. Usambara Mts., Mtai, *Holst* 2460 (B!, syn., G!, P!, isosyn.) & 2466 (B!, syn., K!, P!, isosyn.)

Well-branched shrub to 1–2 m.; twigs soon terete, rusty tomentose with short coarse dendritic hairs. Leaves opposite to subopposite; petiole 1–5 mm. long; lamina coriaceous, lanceolate to elliptic or slightly oblanceolate, 2.5–5.5 cm. long, 1–3.2 cm. wide, bluntly pointed to rounded, basally cuneate to shortly cordate, tomentose beneath with pale sessile stellate hairs and conspicuous rusty dendritic hairs, glabrescent above; lateral nerves 5–7 pairs, spreading, sometimes obscured. Racemes axillary, 2–5 cm. long, 16–14-flowered; pedicel 0.5–1(–2 in fruit) mm. long; bract boat-shaped, 2–3 mm. long. Receptacle 2 mm. long, densely hairy; calyx 0.5 mm. long. Corolla orange-yellow, sometimes redder at extremities, 2.5–3 cm. long, rusty tomentose with short dendritic hairs; tube 8–10 mm. long; basal swelling ovoid, 2.5–3 mm. long; lobes linear-lanceolate over the anthers in bud, linear below, irregularly erect to spreading and reflexed from point of filament-insertion, sometimes coherent to one side. Stamen-filaments slender, inserted 7–8 mm. above the base of the lobes, the upper half slightly thickened, loosely coiling, collapsing and mostly breaking off after anthesis; anthers 2.5–3 mm. long; connective-appendage small, blunt. Berry bluish green, bottle-shaped, 7 mm. long, 4.5 mm. in diameter, hairy, but ultimately glabrescent; seed bright orange. Fig. 15/12–19 (p. 88).

TANZANIA. Lushoto District: Shagayu Forest, Sunga, 3 km. E. of sawmill, 13 July 1983, *Polhill, Lovett & Ruffo* 4994! & 2.5 km. NE. of Bumbuli Mission on path to Mazumbai, 10 May 1953, *Drummond & Hemsley* 2466! & Magamba Forest Reserve, 27 Feb. 1963, *Semsei* 3626!
DISTR. **T** 3; known only from the W. Usambara Mts.
HAB. At edges of drier montane forest and commonly in heath zone above with *Agauria* and *Erica* (*Phillipia*); 1600–2300 m.

SYN. *Loranthus erectus* Engl. in E.J. 20: 99 (1894) & P.O.A. C: 166, t. 16A–C (1895); Sprague in F.T.A. 6(1): 305 (1910); F.D.O.-A. 2: 164 (1932); T.T.C.L.: 281 (1949)
  *Dendrophthoë erecta* (Engl.) Danser in Verh. K. Akad. Wet., sect. 2, 29(6): 44 (1933)

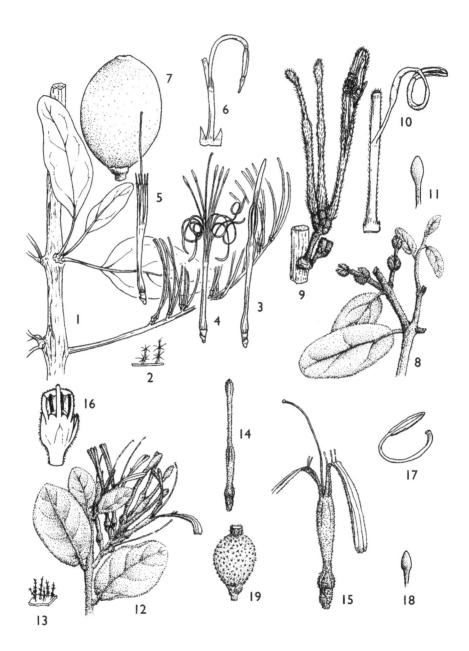

FIG. 15. *OEDINA PENDENS* – **1**, flowering branch, × ²/₃; **2**, hairs, × 20; **3**, flower bud, × 1; **4**, flower, just opened artificially, × 1; **5**, old flower, with parts shed, × 1; **6**, stamen, showing attachment, × 2; **7**, fruit, × 3. *OEDINA CONGDONIANA* — **8**, branchlet with young inflorescences, × 1; **9**, flowering node, × 1; **10**, stamen, showing attachment, × 4; **11**, style-tip, × 4. *OEDINA ERECTA* — **12**, flowering branch, × ²/₃; **13**, hairs, × 10; **14**, flower bud, × 1; **15**, open flower, × 2; **16**, section of base of flower, × 6; **17**, upper part of stamen, × 6; **18**, style-tip, × 6; **19**, fruit, × 2. 1–3, from *Wiens* 6571; 4–7, from *Drummond & Hemsley* 1626; 8, 10, 11, from *Brummitt et al.* 16254; 9, from *Congdon* 139; 12–19, from *Polhill et al.* 4998. Drawn by Christine Grey-Wilson.

3. **O. brevispicata** *Polh. & Wiens*, Mistletoes Afr.: 228, fig. 24D (1998). Type: Tanzania, Morogoro District, Uluguru Mts., near Morogoro, *Schlieben* 3285 (BR!, holo., B!, K!, LISC!, PRE!, SRGH!, iso.)

Shrub to 1 m. or so, well branched; twigs tomentellous with stellate and small rusty dendritic hairs, soon glabrescent. Petiole 5–12 mm. long; lamina thinly coriaceous, oblong-elliptic to oblong-obovate, 3–8 cm. long, 1.5–4 cm. wide, rounded at the apex, cuneate at the base, inconspicuously stellate-pubescent beneath, ± glabrescent; lateral nerves 4–8 pairs, spreading-ascending, rather obscure. Spikes axillary and clustered at older nodes, with 6–12 sessile flowers on an axis 4–25 mm. long; bract ovate to elliptic obovate, concave, 8 1 mm. long, pubescent. Receptacle 1 1.5 mm. long, pubescent; calyx tubular, 2–3 mm. long, ciliate. Corolla 3.5–4 cm. long, orange, stellate-pubescent; bud-tip only slightly swollen; tube 15 mm. long, with a narrowly ellipsoid basal swelling 3–4 mm. long; lobes erect, the upper part spreading to reflexed. Filaments inserted 6–7 mm. above the base of the lobes, the upper half slightly thickened and loosely coiled, sometimes breaking off; anther 3–4.5 mm. long; connective-appendage minute. Berry not seen.

TANZANIA. Lushoto District: Magamba Forest Reserve, 15 km. N. of Magamba, 19 Mar. 1972, *Wiens* 4592!; Morogoro District: Uluguru Mts., near Morogoro, 22 Jan. 1933, *Schlieben* 3285!
DISTR. **T** 3, 6; known only from the above gatherings
HAB. Montane forest; 1400–1900 m.

4. **O. congdoniana** *Polh. & Wiens* in Novon 7: 277, fig. 1R–W (1997) & Mistletoes Afr.: 228, photo. 100, fig. 24B (1998). Type: Malawi, Mafinga Mts., *Brummitt, Polhill & Banda* 16254 (K!, holo., BR!, MAL!, MO!, iso.)

Shrub, spreading, well branched, to 1 m. or so; twigs rusty tomentose with short dendritic hairs. Petiole 2–5 mm. long; lamina thinly coriaceous, reddish green above, oblong-elliptic to oblong-obovate, 3–6 cm. long, 2–3 cm. wide, rounded at the apex, shortly cordate at the base, rusty tomentellous beneath; lateral nerves 6–8 pairs, sometimes obscure. Spikes axillary and at older nodes, with 4–6 paired flowers on an axis 7–15 mm. long; bract ovate to obovate, concave, 2.5–4 mm. long. Receptacle 3 mm. long, hirsute; calyx tubular, 3–4 mm. long, ciliate. Corolla 4–4.5 cm. long, dull yellow beneath a covering of rusty hairs, with a band of red at slits in late bud; hairs mostly subsimple, 2–3 mm. long; bud-tip scarcely swollen; tube 16–19 mm. long, with a short V-slit, the lobes ± coherent opposite; basal swelling slight; lobes erect to weakly spreading, linear, the upper part only slightly expanded. Filaments pale yellow, inserted ± 6 mm. above the base of the lobes, articulate ± two-thirds of the way up, the upper part slightly thickened, coiled at anthesis and tending to break off; anthers 2.5 mm. long; connective-appendage minute. Berry not seen. Fig. 15/8–11.

TANZANIA. Kilosa District: Chonwe Mts., 9.5 km. W. of Kidodi, Nov. 1968, *Akeroyd & Mayuga* 4!; Iringa District: Uzungwa Mts., Udekwa village, forest block to E. of W. Kilombero Forest Reserve, Dec. 1981, *Rodgers & Hall* 2309!; Njombe District: Manda, 11 Jan. 1987, *Congdon* 139!
DISTR. **T** 6, 7; N. Malawi
HAB. Montane forest; 1500–2200 m.

## 14. ONCELLA

Tiegh. in Bull. Soc. Bot. Fr. 42: 251 (1895); Polh. & Wiens, Mistletoes Afr.: 230 (1998)

*Loranthus* L. sect. *Dendrophthoë* (Mart.) Engl. group *Ambigui* Engl. in E.J. 20: 98 (1894)
*L.* sect. *Ambigui* (Engl.) Sprague in F.T.A. 6(1): 262 (1910)

Small well-branched shrubs from a single haustorial attachment; hairs stellate and dendritic; twigs slightly compressed to angular, soon terete. Leaves opposite or subopposite, shortly petiolate, penninerved. Flowers in racemes; bract small, unilateral. Calyx short, entire to shortly toothed. Corolla joined halfway, 5-lobed, radially symmetrical, orange to pink or red, with a dark glandular patch at the base of each lobe, stellate-pubescent, weakly explosive at anthesis; tube with a basal swelling; bud-tip fusiform to obovoid; lobes usually erect, often partly cohering at tips, some or all sometimes spreading to slightly reflexed from near the point of filament-insertion, occasionally coiled and tending to break off. Filaments attached a little less than halfway up the corolla-lobes, slender in lower persistent part, jointed and slightly thickened above, the upper part coiling, collapsing and breaking off to various degrees; anthers basifixed, 4-thecous; if the thecae short (less than 2 mm.) then unequal, the dorsal ones shorter; connective-appendage minute to subulate or expanded and ± bilobed. Style filiform; stigma ovoid to obovoid or fusiform. Berry white to red (sometimes turning black), ellipsoid, with persistent calyx.

4 species in eastern Africa, montane and coastal.

A small group of similar species closely related to *Oedina* and *Erianthemum*. The flowers are regular and characterised by the glandular patches at the base of the corolla-lobes and the generally club-shaped heads of the buds enclosing the exceptionally small anthers (longer in *O. schliebeniana*). The flowers are opened as in *Oedina*.

1. Anthers 2–3 mm. long, the thecae subequal; tip of bud fusiform over the anthers; corolla-lobes ± spreading at anthesis, more than 1.5 times as long as the tube · · · · ·    1. *O. schliebeniana*
   Anthers 1.2–1.8 mm. long, the dorsal thecae shorter; tip of bud clavate, obovoid over the anthers; corolla-lobes remaining erect, as long as or a little longer than the tube · · · · · · · · · · · · · · · · · · · · · · · · · · · · · · · · · · · · · · · · 2
2. Flowers 2–8 on axes up to 1.5 cm. long; leaves 1.5–4 × 1–2.5 cm. · · · · · · · · · · · · · · · · · · · · · · · · · · · · · · · · · · ·    4. *O. gracilis*
   Flowers 12–40 on 2–7 cm. long axes; leaves mostly 3.5–10 × 1.5–6 cm. · · · · · · · · · · · · · · · · · · · · · · · · · · · · · · · · · · · · 3
3. Pedicels 2–4 mm. long; corolla apricot-yellow to orange, inconspicuously puberulous; leaves broadest about the middle or below · · · · · · · · · · · · · · · · · · · · · · · · ·    2. *O. ambigua*
   Pedicels 0.5–1 mm. long; corolla pinkish, with densely grey-white puberulous tips to buds; leaves mostly broadest above the middle · · · · · · · · · · · · · · · · · · · · · · · · ·    3. *O. curviramea*

1. **O. schliebeniana** *Polh. & Wiens*, Mistletoes Afr.: 230 (1998) Type: Tanzania, Lindi District, Lake Lutamba, *Schlieben* 6121 (BR!, holo., B!, BRLU!, K!, LISC!, PRE!, iso.)

Small shrub with relatively short spreading branchlets; twigs soon terete, tomentellous with short rusty dendritic hairs, soon glabrescent. Petiole 3–5 mm. long; lamina thinly coriaceous, elliptic to obovate, 2–4 cm. long, 1–2 cm. wide, obtuse or rounded at the apex, cuneate to shortly rounded at the base, rusty tomentellous beneath but soon glabrescent apart from a few scattered dendritic hairs, with 4–8 pairs of lateral nerves, the lower ones sometimes rather strongly ascending. Racemes 1–several in axils and at older nodes, with 4–8 flowers on an axis 1–2 cm. long; pedicels 1–2 mm. long; bract boat-shaped, 1.5–2 mm. long. Receptacle 1–1.5 mm. long, puberulous; calyx 0.7–0.8 mm. long, slightly toothed, ciliate. Corolla regular, 3.8–4.5 cm. long, orange-red to carmine, puberulous, glabrescent, especially on the lobes; bud-tip fusiform; tube 1.5–1.8 cm. long; basal swelling ovoid to globose, 3 mm. long, tube narrow for 4–5 mm. above, then flared; lobes erect, with the upper half reflexing. Filaments inserted 6–8 mm. above the base of the lobes, the upper two-thirds slightly thickened, loosely coiled and generally breaking off; anthers 2–3 mm. long, with a minute connective-appendage. Berry not seen.

TANZANIA. Lindi District: Rondo Plateau, 22 Feb. 1987, *Congdon* 144! & E. side of Lake
Lutamba, Litipo Forest Reserve, 26 Feb. 1991, *Bidgood, Abdallah & Vollesen* 1727!; Newala, 6
Mar. 1959, *Hay* 43!
DISTR. **T** 8; not known elsewhere
HAB. Coastal bushland and thicket, recorded on *Hymenocardia*; 150–850 m.

NOTE. The elongate anthers with a small connective have not diverged from the arrangement
in *Oedina*, but the corolla, with characteristic glandular patches at the base of the lobes, and
the general appearance of the plant fit better with *Oncella.*

2. **O. ambigua** (*Engl.*) *Tiegh.* in Bull. Soc. Bot. Fr. 42. 251 (1895), Danser in Verh.
K. Akad. Wet., sect. 2, 29(6): 97 (1933); Polh. & Wiens, Mistletoes Afr.: 231, photo.
101 (1998). Type: Tanzania, Bagamoyo District, Wami R., *Hildebrandt* 1032 (B!, holo.,
BM!, K!, P!, iso.)

Small shrub to 1 m. or so; twigs stellate-puberulous, soon glabrescent. Petiole
0.5–7 mm. long; lamina pale rather yellow green to full green, elliptic-lanceolate to
elliptic or ovate-elliptic, 3.5–10 cm. long, 1.5–6 cm. wide, bluntly pointed to
subacute at the apex, basally cordate to rounded, soon glabrescent; lateral nerves
4–8 pairs. Racemes axillary and at older nodes, 12–24-flowered; axis 2–4 cm. long,
whitish puberulous; pedicel 2–4 mm. long; bract triangular-ovate, 0.8–1.5 mm.
long. Receptacle 1–1.5 mm. long; calyx very short. Corolla 2.5–3.8 cm. long,
apricot-yellow to orange with a dark spot at the base of each lobe, and sometimes
a greenish basal swelling, thinly stellate puberulous; bud-tip clavate; tube 1.3–1.8
cm. long; basal swelling (2–)3–4 mm. long, narrow for 1.5–2 mm. above and then
flared; lobes erect, ± cohering at the expanded tips. Filaments inserted 5–6 mm.
above the base of the lobes, the upper part thickened, coiled and generally
breaking off; anthers 1.2–1.8 mm. long, the dorsal thecae relatively short, with a
broad ± slightly bilobed connective-appendage. Berry red turning black, ellipsoid,
5–8 mm. long, 4–5 mm. in diameter.

KENYA. Tana R. District: 1 km. S. of Garsen, W. of Tana R., 15 July 1972, *Gillett & Kibuwa* 19913!
& Tana River National Primate Reserve, Mchelelo Forest, 11 Mar. 1990, *Luke et al.* in T.P.R.
68!; Lamu District: Mambosasa, 1929, *R.M. Graham* in F.D. 1799!
TANZANIA. Tanga District: Kigombe, Sept. 1955, *Semsei* 2339!; Bagamoyo District: 30 km. NW. of
Dar es Salaam off road to Bagamoyo, 30 Apr. 1975, *Hepper & Field* 5200!; Uzaramo District:
Kunduchi salt works, 31 July 1969, *B.J. Harris & McCusker* 3026!; Zanzibar I., Kiwengwa, 13
Mar. 1962, *Faulkner* 3019!
DISTR. **K** 7; **T** 3, 6; **Z**; not known elsewhere
HAB. Coastal bushland and riverine forest, on various hosts, including *Antidesma, Grewia,
Securinega* and *Trichilia*; 0–300 m.

SYN. *Loranthus ambiguus* Engl. in E.J. 20: 98 (1894) & P.O.A. C: 166, t. 18D–G (1895); Sprague
in F.T.A. 6(1): 307 (1910); F.D.O.-A. 2: 164 (1932); T.T.C.L.: 280 (1949)
*L. ambiguus* Engl. var. *subacutus* Engl. in E.J. 20: 99 (1894) & P.O.A. C: 166 (1895); Sprague
in F.T.A. 6(1): 307 (1910); F.D.O.-A. 2: 164 (1932); T.T.C.L.: 280 (1949). Types:
Tanzania, Bagamoyo, *Stuhlmann* 7 & Dunda, *Stuhlmann* 6510 & Uzaramo District, Dar es
Salaam, *Stuhlmann* 7316 (B!, syn.)
*Oncella sacleuxii* Tiegh. in Bull. Soc. Bot. Fr. 42: 251 (1895); Danser in Verh. K. Akad. Wet.,
sect. 2, 29(6): 97 (1933). Type: Tanzania, Bagamoyo, *Sacleux* 676 (P!, holo.)
*Loranthus sacleuxii* (Tiegh.) Engl. in E. & P. Pf., Nachtr. 1: 131 (1897); Sprague in F.T.A.
6(1): 307 (1910); T.T.C.L.: 281 (1949)
*L. poecilobotrys* Werth, Veg. Ins. Sansibar: 35 (1901); Sprague in F.T.A. 6(1): 306 (1910);
F.D.O.-A. 2: 164 (1932); T.T.C.L.: 281 (1949); U.O.P.Z.: 335 (1949). Type: Tanzania,
Zanzibar I., Mitschawi, *Werth* (B!, holo., K!, fragment)
*Oncella poecilobotrys* (Werth) Danser in Verh. K. Akad. Wet., sect. 2, 29(6): 97 (1933)

NOTE. The populations from Zanzibar, described as *O. poecilobotrys*, have notably short petioles,
0.5–1(–2) mm. long (never less than 2 mm. on the mainland), but in other respects no
significant differences are apparent.

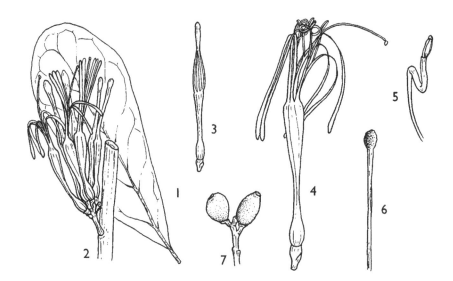

FIG. 16. *ONCELLA CURVIRAMEA* — **1**, leaf, × 1; **2**, inflorescence, × 1; **3**, flower bud, × 1; **4**, open flower, × 2; **5**, stamen, × 6; **6**, style-tip, × 6; **7**, fruits, × 2. 1, from *Polhill et al.* 4972; 2–6, from *Drummond & Hemsley* 3948; 7, from *Greenway* 5028. Drawn by Christine Grey-Wilson.

3. **O. curviramea** (*Engl.*) *Danser* in Verh. K. Akad. Wet., sect. 2, 29(6): 97 (1933); Polh. & Wiens, Mistletoes Afr.: 232, photo. 102, fig. 25A (1998). Type: Tanzania, Uzaramo District, Vikindu, *Stuhlmann* 6107 (B!, holo.)

Shrub to 0.5–2.5 m., becoming pendent; twigs stellate-puberulous, soon glabrescent. Petiole 5–10 mm. long; lamina elliptic-oblanceolate to elliptic-obovate or obovate-circular, 4–10 cm. long, 1.5–5 cm. wide, bluntly pointed to rounded at the apex, cuneate to shortly rounded at the base, soon glabrescent; lateral nerves 2–4 pairs. Racemes terminal on very short axillary shoots (bearing 2 leaves) or at old nodes, 16–40-flowered; axis 3–7 cm. long, densely whitish hairy; pedicels 0.5–1 mm. long, extended upwards by an ovate-triangular bract of same length. Receptacle 1.5 mm. long; calyx 0.3–0.5 mm. long. Corolla 3–3.7 cm. long, pinkish red, sometimes green at the base, with dark spots at the base of the lobes, and with fairly dense pale hairs covering the upper part of the lobes; bud-tip clavate; tube 1.5–1.8 cm. long; basal swelling 2.5–4 mm. long, narrowed for 2–3 mm. and then flared; lobes erect, recurved at the tip. Stamens similar to *O. ambigua*. Berry pink to white, ellipsoid, 7 mm. long, 4–5 mm. in diameter. Fig. 16.

KENYA. Kwale District: Waa, 27 Sept. 1982, *Polhill & S.A. Robertson* 4846!; Mombasa District: Bamburi to Shimo la Tewa, 24 Nov. 1971, *Bally & Smith* 14379!; Kilifi District: Malindi, 3 Jan. 1994, *S.A. Robertson* 6880!
TANZANIA. Tanga District: E. Usambara Mts., Lutindi Forest Reserve, SE. part, 16 Nov. 1986, *Iversen, Steiner & Temu* 86799!; Rufiji District: Mafia, Miewi Mdogo, 5 Sept. 1937, *Greenway* 5227!; Lindi District: Rondo Forest Reserve, 6 km. N. of Forest Station, 11 Feb. 1991, *Bidgood, Abdallah & Vollesen* 1492!
DISTR. **K** 7; **T** 3, 6, 8; Mozambique
HAB. Coastal forest and bushland, extending up rivers to montane forest in Usambara Mts., on a wide variety of hosts; 0–2050 m.

SYN. *Loranthus curvirameus* Engl., P.O.A. C: 165 (1895) & in E.J. 40: 523 (1908); Sprague in F.T.A. 6(1): 306 (1910); F.D.O.-A. 2: 164 (1932); T.T.C.L.: 281 (1949)

4. **O. gracilis** *Polh. & Wiens*, Mistletoes Afr.: 233, photo. 103 (1998). Type: Tanzania, Uluguru Mts., near Morogoro, *Schlieben* 4012 (BR!, holo., B!, LISC!, PRE!, iso.)

Small much-branched shrub to 50 cm.; twigs stellate-puberulous, very soon glabrescent. Petiole 2–5 mm. long; lamina thinly coriaceous, dull green, elliptic to elliptic-obovate, 1.5–4 cm. long, 1–2.5 cm. wide, rounded at the apex, cuneate at the base, soon glabrous; lateral nerves 2–6 pairs, often obscure. Racemes axillary and at older nodes, with 2–8 flowers (often rather crowded above if few); axis 3–15 mm. long, puberulous; pedicels 1–1.5 mm. long; bract narrowly ovate-triangular to lanceolate, 1–2 mm. long. Receptacle 1–1.5 mm. long, puberulous; calyx 0.3–0.5 mm. long. Corolla 3–3.5 cm. long, pinkish red, with dark marks at the base of the lobes, thinly stellate-puberulous; bud-tip clavate; tube 15 mm. long, with basal swelling ovoid-ellipsoid, 2–3 mm. long, narrowed for 2–4 mm., then gradually flared; lobes erect, expanded at the tips. Filaments inserted 5 mm. above the base of the lobes, the upper part thickened, coiling and detaching; anthers 1.5–1.8 mm. long, the dorsal thecae relatively short, with a relatively conspicuous truncate connective-appendage. Berry not seen.

TANZANIA. Morogoro District: Uluguru Mts., near Morogoro, 21 June 1933, *Schlieben* 4012! & Lupanga Peak, 26 June 1983, *Polhill & Lovett* 4919! & above Mgeta, 3 Jan. 1986, *D. & C. Wiens & Pócs* 6535!
DISTR. **T** 6; known only from the Uluguru Mts.
HAB. Montane forest; 1700–1800 m.

NOTE. Most similar to *O. curviramea*, but altogether much more delicate and with less hairy flowers.


## 15. **ERIANTHEMUM**

Tiegh. in Bull. Soc. Bot. Fr. 42: 247 (1895); Polh. & Wiens, Mistletoes Afr.: 234 (1998)

*Loranthus* L. sect. *Dendrophthoë* (Mart.) Engl. group *Hirsuti* Engl. in E.J. 20: 104 (1894)
*L.* sect. *Hirsuti* (Engl.) Sprague in F.T.A. 6(1): 263 (1910)

Shrubs mostly 1–2 m. from a single haustorial attachment; hairs stellate and dendritic, longer ones on flowers subsimple; twigs terete. Leaves opposite, subopposite or partly alternate, sometimes clustered on short shoots, penninerved. Flowers generally few, sessile in heads, elsewhere sometimes shortly racemose or spicate; peduncles axillary, clustered at older nodes or terminal short shoots bearing (1–)2 pairs of leaves; bract unilateral, ovate to subcircular, concave, sometimes with a leafy limb. Receptacle generally long-hairy; calyx cupular to tubular, subentire to shortly toothed. Corolla usually joined less than halfway, 5-lobed, radially symmetrical, mostly yellow to green, flushed orange on tube, silky villous or at least with stellate and some long subsimple hairs on buds; tube constricted above a basal swelling, glabrous or hairy inside; lobes spreading-reflexed from below the middle. Filaments attached well above the lobe-base, the upper thickened and flattened part breaking off with anthers when flower is opened; anthers 4-thecous, with a small bifid or hammer-shaped apical connective-appendage. Style filiform; stigma capitate. Berry orange, reddish or blue-green, with persistent calyx urceolate; seed orange to red.

16 species in eastern and southern Africa from Ethiopia to the Eastern Cape Province of South Africa, extending westwards to the Central African Republic, Shaba and Angola.

Closely related to *Oncella* and with some overlap of distinguishing features, but forming a coherent group that is soon recognised by familiarity with the common and widespread species, *E. dregei*.
Vents appear in the mature flower-buds through which sunbirds seek nectar, the withdrawal of the beak tearing between the corolla-lobes, releasing the tension that explosively sheds the anthers with the coiling upper part of the filament and deposits a puff of pollen on the bird's head.

1. Peduncles axillary or clustered at nodes below; leaves
crowded only at base of long shoots · · · · · · · · · · · · · · · · · · · · · · · · · · · 2
Peduncles terminal on leafy short shoots, sometimes also
axillary on long shoots (beware fragmentary gatherings
of *E. nyikense* and *E. lindense* in particular) · · · · · · · · · · · · · · · · · · 4
2. Calyx plus receptacle 8–10 mm. long; corolla 5–6 cm. long,
densely covered with long distinctly yellowish hairs;
leaf-lamina (5–)9–15 × (3–)4.5–12 cm., with 7–10 pairs
of lateral nerves · · · · · · · · · · · · · · · · · · · · · · · · · · 3. *E. taborense*
Calyx plus receptacle 3.5–7 mm. long; corolla 3–5(–5.5)
cm. long, variously clothed with long whitish, cream,
brownish or rarely yellowish hairs; leaf-lamina 3–10.5 ×
1.2–7 cm., with 4–8 pairs of lateral nerves · · · · · · · · · · · · · · · · · · · · · 3
3. Leaves subglabrous above, finely brownish stellate
tomentellous beneath; peduncles mostly below current
leaves, 2–7(–9) mm. long; long corolla-hairs well
developed on upper part of basal swelling, often sparse
or shorter on tube above at maturity · · · · · · · · · · · · · 1. *E. schelei*
Leaves not so discolorous; peduncles among leaves and
below, 0–18 mm. long; corolla-indumentum variable,
sometimes as above (especially near coast), often more
uniformly dense · · · · · · · · · · · · · · · · · · · · · · · · · · 2. *E. dregei*
4. Constriction of corolla-tube (2.5–)3–5 mm. long, glabrous
inside; receptacle plus calyx 3–7 mm. long, the calyx
distinctly longer than the receptacle; berry pinkish-red
where known · · · · · · · · · · · · · · · · · · · · · · · · · · · · · · · · · · · · · · · · 5
Constriction of corolla-tube 1–2 mm. long, hairy inside
(fig. 17/7); receptacle plus calyx 1.5–3(–3.5) mm. long,
the calyx shorter to a little longer than the receptacle;
berry bluish where known · · · · · · · · · · · · · · · · · · · · · · · · · · · · · · · 10
5. Bracts ovate, concave, the inner ones at least without a
leafy limb, the outer ones sometimes with a small limb · · · · · · · · · · · · · · 6
Bracts all with a well-developed leafy limb · · · · · · · · · · · · · · · · · · · · · 8
6. Leaves sessile, subcircular, glabrous · · · · · · · · · · · · · · · 6. *E. rotundifolium*
Leaves petiolate, elliptic to ovate or obovate, hairy · · · · · · · · · · · · · · · · · 7
7. Leaf-lamina 1.5–3 times as long as broad, felty with a fine
woolly tomentum completely covering the surface · · · · 4. *E. lanatum*
Leaf-lamina 5–6 times as long as broad on long shoots
(sometimes less on short shoots), stellate-tomentose at
first but soon glabrescent · · · · · · · · · · · · · · · · · · · · · 5. *E. schmitzii*
8. Leaf-blades all cuneate to shortly rounded at base,
glabrescent; calyx 2.5–3 mm. long, villous; peduncles
1–2.5 cm. long · · · · · · · · · · · · · · · · · · · · · · · · · · · · · 9. *E. alveatum*
Leaf-blades mostly shortly cordate at base (except
youngest on short shoots), glabrescent or rather
persistently hairy · · · · · · · · · · · · · · · · · · · · · · · · · · · · · · · · · · · · · 9
9. Calyx 2–3 mm. long, villous; peduncles 0.3–2.5(–3) cm.
long; berry pilose · · · · · · · · · · · · · · · · · · · · · · · · · · · 7. *E. nyikense*
Calyx 3–4.5 mm. long, glabrous or nearly so except at
margin; peduncles (1–)2–4 cm. long; berry subglabrous 8. *E. lindense*
10. Corolla extensively hairy; bracts up to 10 mm. long, without
or with a small leafy limb · · · · · · · · · · · · · · · · · · · · · 12. *E. aethiopicum*
Corolla sparsely hairy at least on lobes; bracts leafy, 10–20
mm. long · · · · · · · · · · · · · · · · · · · · · · · · · · · · · · · · · · · · · · · · · · 11

11. Corolla-tube hairy; petiole 1–2 mm. long; bracts broadest
    about the middle · · · · · · · · · · · · · · · · · · · · · · · · ·        10. *E. commiphorae*
    Corolla-tube with scattered hairs only; petiole 2–5 mm.
    long; bracts broadest above middle · · · · · · · · · · · · ·        11. *E. occultum*

1. **E. schelei** (*Engl.*) *Tiegh.* in Bull. Soc. Bot. Fr. 42: 248 (1895); Danser in Verh. K. Akad. Wet., sect. 2, 29(6): 54 (1933); Polh. & Wiens, Mistletoes Afr.: 236, photo. 104 (1998). Type: Tanzania, Lushoto District, E. Usambara Mts., Lutindi, *Holst* 3302 (B!, holo., K!, P!, US!, iso.)

Stems spreading or pendent to 1 m. or so, twigs tomentose with fulvous stellate and shortly dendritic hairs, glabrescent. Leaf-pairs well spaced; petiole 5–15 mm. long; lamina broadly oblong-elliptic to ovate or subcircular, 3–10.5 cm. long, 3.5–7 cm. wide, cordate at the base, glabrous or nearly so above, fulvous stellate-tomentellous beneath, with 4–7 pairs of obscure lateral nerves. Heads mostly clustered at old nodes, a few sometimes axillary, (2–)4-flowered; peduncle 2–7(–9) mm. long, tomentose; bract ovate, concave, 3–5 mm. long, occasionally with a small leafy limb, then to 10 mm. long, thinly to densely stellate pubescent. Receptacle 1–1.5 mm. long, long-hairy; calyx 3–4 mm. long, often slightly urceolate, sparsely to generally densely silky villous. Corolla 4–5 cm. long, with basal swelling green, rest of tube orange, lobes pale green with red markings at their base, white silky villous over shorter branched hairs, the simple hairs notably long on upper part of basal swelling and usually rather sparse on rest of tube; basal swelling 4–7 mm. long; constriction (3–)4–7mm. long; funnel-shaped upper tube 7–9 mm. long. Berry orange, 10 mm. long including calyx, subglabrous to thinly pilose; seed red.

TANZANIA. Morogoro District: Uluguru Mts., Mgeta R. valley above Bunduki, 12 Mar. 1953, *Drummond & Hemsley* 1487!; Iringa District: Ihangana Forest Reserve, near Kibengu, 14 Feb. 1962, *Polhill & Paulo* 1473!; Songea District: 3 km. E. of Ndengo, 13 Jan. 1956, *Milne-Redhead & Taylor* 8323!
DISTR. **T** 3, 6–8; Malawi
HAB. Montane and riverine forests, on a wide variety of hosts; 900–1950 m.

SYN. *Loranthus schelei* Engl., Abh. Preuss. Akad. Wiss.: 53 (1894) & in E.J. 20: 105 (1894) & P.O.A. C: 166, t. 17E–J (1895); Sprague in F.T.A. 6(1): 310 (1910); F.D.O.-A. 2: 165 (1932), pro parte; T.T.C.L.: 283 (1949)
        *L. ulugurensis* Engl. in E.J. 28: 381 (1900); Sprague in F.T.A. 6(1): 309 (1910); F.D.O.-A. 2: 165 (1932); T.T.C.L.: 283 (1949). Type: Tanzania, Morogoro District, Uluguru Mts., Nglewenu [Ngthweme], *Stuhlmann* 8832 (B!, holo., K!, fragment)
        [*L. dregei* Eckl. & Zeyh. forma *obtusifolius* sensu Engl. in E.J. 28: 383 (1900), pro parte quoad *Goetze* 162, *non* Engl. (1894)]
        *Erianthemum ulugurense* (Engl.) Danser in Verh. K. Akad. Wet., sect. 2, 29(6): 54 (1933)

NOTE. *E. schelei* is very similar to *E. dregei* but the disjunct forest populations are relatively uniform and in minor features consistently different from that complex. The only apparent deviation is in the western and southern Uluguru Mts., where some specimens (including the type of *L. ulugurensis*) are aberrant with a tendency to form foliaceous bracts and to have fewer hairs on the calyx and sometimes the corolla. Both forms grow on the Mgeta R. above Bunduki and it is just possible that two species are involved.

2. **E. dregei** (*Eckl. & Zeyh.*) *Tiegh.* in Bull. Soc. Bot. Fr. 42: 248 (1895); Danser in Verh. K. Akad. Wet., sect. 2, 29(6): 53 (1933); Wiens & Tölken in F.S.A. 10: 30, fig. 12/1 (1979); M.G. Gilbert in Fl. Ethiopia 3: 366, fig. 114.7/6 (1990); Germish. et al. in Fl. Pl. Afr. 53, t. 2084 (1994); Polh. & Wiens in U.K.W.F., ed. 2: 157, t. 56 (1994) & Mistletoes Afr.: 237, photo. 7, 105, fig. 26A (1998). Type: South Africa, Eastern Cape Province, Bothasberg, *Ecklon & Zeyher* 2284 (B!, G!, MO!, iso.)

Stems spreading to pendent, well branched, to 1.5 m. or so; twigs tomentellous with white to fulvous stellate and shortly dendritic hairs, glabrescent. Leaves opposite to alternate, well spaced except at the base of new shoots; petiole 3–15 mm. long; lamina elliptic-oblong to ovate, 3–10 cm. long, 1.2–6 cm. wide, cuneate, rounded or cordate at the base, glabrous to subdensely stellate-pubescent, glabrescent, with 4–8 pairs of lateral nerves. Heads 1–several in axils or at nodes below, 2–6-flowered; peduncle 0–18 mm. long; bract broadly ovate, concave, 2–5 mm. long, sometimes with a leafy limb and up to 15 mm. long, stellate-pubescent. Receptacle 1–1.5 mm. long, long-hairy; calyx tubular, 2.5–6 mm. long, thinly to densely silky villous. Corolla 3–5(–5.5) cm. long, pale often yellowish green, generally flushed orange to pink on the tube above the basal swelling, usually densely silky villous, the long whitish, cream or sometimes reddish hairs masking the constriction, but sometimes (especially in coastal plants) long hairs rather sparse on the basal part of the then notably waisted tube; basal swelling 3–7 mm. long; constriction 3–6 mm. long; funnel-shaped upper tube 4–12 mm. long. Berry orange to bright red, 10–15 mm. long including the calyx, pilose; seed red. Fig. 17/1–4.

UGANDA. Karamoja District: Timu Forest, 6 Nov. 1939, *A.S. Thomas* 3205! & Kamion, 8 Nov. 1939, *A.S. Thomas* 3246!; Mbale District: Elgon, Kapchorwa, 8 Sept. 1954, *Lind* 276!
KENYA. Northern Frontier Province: Moyale, 28 Apr. 1952, *Gillett* 12952!; Machakos District: Lukenya, 24 Sept. 1978, *M.G. Gilbert* 5059!; Kilifi District: 9 km. S. of Malindi on Mombasa road, 16 Jan. 1972, *Wiens* 4527!
TANZANIA. Mbulu District: Lake Manyara National Park, near Msasa, 28 Feb. 1964, *Greenway & Kanuri* 11261!; Handeni District: Kideleko Mission, 16 July 1957, *Semsei* 2668!; Tunduru District: ± 11 km. E. of Songea District boundary, 7 June 1956, *Milne-Redhead & Taylor* 10680!; Zanzibar I., Chukwani, 7 Aug. 1959, *Faulkner* 2325!
DISTR. U 1, 3; K 1–4, 6, 7; T 1–3, 5–8; Z; eastern Africa from N. Ethiopia south to the Eastern Cape Province of South Africa, extending westwards to S. Angola
HAB. Widely distributed from forest edges to woodland, bushland, wooded grassland and disturbed places, on many different hosts; 0–2650 m.

SYN. *Loranthus dregei* Eckl. & Zeyh., Enum. Pl. Afr. Austr.: 358 (1837); Harv. in Fl. Cap. 2: 575 (1862); Engl. in E.J. 20: 104 (1894) & P.O.A. C: 166 (1895); Sprague in F.T.A. 6(1): 311 (1910), pro parte, & in Fl. Cap. 5(2): 109 (1915); F.D.O.-A. 2: 165 (1932); T.T.C.L.: 281 (1949), pro parte; U.O.P.Z.: 335 (1949)
    *L. roseus* Klotzsch in Peters, Reise Mossamb. Bot. 1: 177 (1861). Type: Mozambique, Maputo, Delagoa Bay, *Peters* (B!, holo.)
    *L. hirsutiflorus* Klotzsch in Peters, Reise Mossamb. Bot. 1: 178 (1861). Type: Mozambique, between Sena and Tete, *Peters* 8 (B!, holo., K!, iso.)
    *L. dregei* Eckl. & Zeyh. forma *subcuneifolius* Engl. in E.J. 20: 104 (1894). Lectotype, chosen by Sprague in F.T.A. 6(1): 312 (1910) & K.B. 1911: 138 (1911): Ethiopia, near Gapdia, *Schimper* II. 768 (B!, lecto., BM!, K!, P!, isolecto.)
    *L. dregei* Eckl. & Zeyh. forma *obtusifolius* Engl. in E.J. 20: 105 (1894) & in E.J. 28: 383 (1900), pro parte, & in E.J. 30: 302 (1901). Type as for *L. roseus* Klotzsch
    *L. dregei* Eckl. & Zeyh. var. *sodenii* Engl. in E.J. 20: 105 (1894) & P.O.A. C: 166 (1895); Sprague in F.T.A. 6(1): 313 (1910); F.D.O.-A. 2: 166 (1932); T.T.C.L.: 282 (1949) & in Mem. N.Y. Bot. Gard. 9: 63 (1954). Lectotype, chosen by Polh. & Wiens (1998): Tanzania, Uzaramo District, Dar es Salaam, *Hildebrandt* 1224 (K!, lecto.)
    *L. dregei* Eckl. & Zeyh. var. *subcuneifolius* (Engl.) Sprague in F.T.A. 6(1): 312 (1910) & in K.B. 1911: 138 (1911); F.D.O.-A. 2: 166 (1932); T.T.C.L.: 282 (1949)
    *L. dregei* Eckl. & Zeyh. var. *nyasicus* Sprague in F.T.A. 6(1): 313 (1910) & K.B. 1911: 139 (1911); F.D.O.-A. 2: 167 (1932); T.T.C.L.: 282 (1949). Lectotype, chosen by Polh. & Wiens (1998): Malawi, Zomba, *Purves* 154 (K!, lecto.)
    *L. dregei* Eckl. & Zeyh. var. *ovatus* Sprague in F.T.A. 6(1): 314 (1910) & in K.B. 1911: 139 (1911); F.D.O.-A. 2: 167 (1932); T.T.C.L.: 282 (1949). Lectotype, chosen by Polh. & Wiens (1998): Tanzania, Lushoto District, E. Usambara Mts., Amani, *Warnecke* 349 (BM!, lecto., EA!, isolecto., K!, fragment)
    *L. dregei* Eckl. & Zeyh. var. *foliaceus* Sprague in F.T.A. 6(1): 314 (1910) & in K.B. 1911: 139 (1911); T.T.C.L.: 281 (1949). Lectotype, chosen by Polh. & Wiens (1998): Kenya, Machakos District, Kibwezi, *Scheffler* 346 (K!, lecto., BM!, P!, WAG!, isolecto.)
    *L. dregei* Eckl. & Zeyh. var. *kilimanjaricus* Sprague in F.T.A. 6(1): 314 (1910) & in K.B. 1911: 140 (1911); K. Krause in N.B.G.B. 8: 499 (1923); F.D.O.-A. 2: 167 (1932); T.T.C.L.: 281

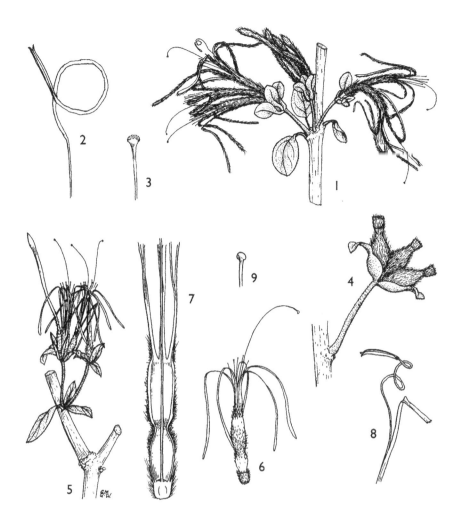

FIG. 17. *ERIANTHEMUM DREGEI* — **1** flowering node, × ²/₃; **2**, stamen, × 4; **3**, style-tip, × 4; **4**, young fruits, × 2. *ERIANTHEMUM COMMIPHORAE* — **5**, flowering branch, × ²/₃; **6**, flower, × 1; **7**, section of flower × 2; **8**, stamen, showing attachment, × 2; **9**, style-tip × 2. 1, from *Wiens* 4543; 2, 3, from *Compton* 28966; 4, from *Ash* 1945; 5–9, from *Polhill & Paulo* 1367. Drawn by Christine Grey-Wilson.

(1949). Type: Tanzania, Moshi District, Kilimanjaro, Marangu, *Volkens* 1934 (K!, holo., B!, BM!, G!, iso.)

L. *heterochromus* K. Krause in N.B.G.B. 8: 500 (1923), *nom. illegit.*, *non* K. Krause (1922). Type: Kenya, N. Nyeri District, Mt. Kenya Forest Station, *R.E. & T.C.E. Fries* 444 (B, holo., BR!, K!, iso.)

L. *linguiformis* Peter, F.D.O.-A. 2: 168, Descr. 14, fig. 13/2 (1932); T.T.C.L.: 282 (1949). Type: Tanzania, Dodoma District, Saranda, *Peter* 44586 (B, holo.)

*Erianthemum heterochromum* Danser in Verh. K. Akad. Wet., sect. 2, 29(6): 54 (1933). Type as for L. *heterochromus* K. Krause

*E. linguiforme* (Peter) Danser in Rec. Trav. Bot. Néerl. 31: 223 (1934)

[*Loranthus ulugurensis* sensu U.K.W.F.: 330, fig. on 335 (1974), *non* Engl.]

*Erianthemum hirsutiflorum* (Klotzsch) Balle, ined.; Vollesen in Opera Bot. 59: 63 (1980), *nom. invalid.*, *comb. non rite publ.*

NOTE. The most common and variable species in much of East Africa. Plants from the coastal region, described as var. *sodenii*, tend to have small flowers, small bracts and more elongate leaves. Over most of the highlands east of the Rift from C. Tanzania to N. Somalia the bracts are generally (not invariably) foliaceous (var. *foliaceus*; *L. heterochromus*) and the leaves tend to be relatively short, cordate and subpersistently hairy. West of the Rift, around Lake Victoria, on Elgon and in NW. Ethiopia, the bracts are less likely to be leafy and the leaves are more variably shaped. In western Tanzania the species is replaced by *E. taborense* and in the *Brachystegia* woodlands *E. dregei* is uncommon and notably variable in the shape of the leaves. The forest element from the Usambara Mts. to Malawi is maintained as a separate species, *E. schelei*.

The name *Loranthus dregei* var. *curvifolius* Engl. is mentioned by Engler in Schinz, Pl. Menyharth.: 409 (1905). Sprague in F.T.A. 6(1): 312 (1910) has interpreted this as just a mistake for forma *subcuneifolius* Engl. There is an observation about the curved flowers, but Sprague does not regard this as a validating description. The specimen, *Menyharth 509*, from Tete Province of Mozambique, has not been seen.

3. **E. taborense** *(Engl.)* *Tiegh.* in Bull. Soc. Bot. Fr. 42: 248 (1895); Polh. & Wiens, Mistletoes Afr.: 238 (1998). Type: Tanzania, Tabora, *Stuhlmann* 573 (B!, holo.)

Stems spreading to pendent, to 1.5–2 m. long; twigs soon thickening to 5–8 mm. diameter, tomentellous with white to fulvous stellate and very shortly dendritic hairs, glabrescent. Leaves opposite or subopposite, well spaced; petiole 5–15 mm. long; lamina ovate to oblong-elliptic, rarely some obovate, (5–)9–15 cm. long, (3–)4.5–12 cm. wide, broadly cuneate to mostly cordate at the base, stellate-puberulous, tardily glabrescent, with 7–10 pairs of lateral nerves. Heads 1–several in axils or at nodes below, 2–6-flowered; peduncle 3–12 mm. long; bract ovate-circular, concave, 4–6 mm. long, pubescent. Receptacle 2 mm. long, long-hairy; calyx tubular, 6–8 mm. long, silky villous. Corolla 5–6 cm. long, pale greenish yellow, flushing pale orange on tube, very densely covered with long silky pale to lemon-yellow hairs masking the constriction of the tube; basal swelling 5–11 mm. long; constriction 4–5 mm. long; distal part of tube ± 10 mm. long. Berry pink to red, 20–23 mm. long including calyx, pilose, ultimately glabrescent; seed orange-yellow, shading to red at the blunt end.

TANZANIA. Mpanda District: Mwesi [Mwese], 20 May 1975, *Kahurananga, Kibuwa & Mungai* 2571!; Mpwapwa, 21 July 1938, *Hornby* 904!; Iringa, just N. of township, 14 July 1956, *Milne-Redhead & Taylor* 11079!
DISTR. **T** 4, 5, 7; Zaire (Shaba), Burundi, Zambia and Malawi
HAB. *Brachystegia* woodland, on a variety of hosts; 800–1650 m.

SYN. *Loranthus taborensis* Engl. in E.J. 20: 106, t. 2D (1894) & P.O.A. C: 166 (1895), excl. fig. cit., & in E.J. 30: 301 (1901)
   *L. dregei Eckl. & Zeyh.* var. *taborensis* (Engl.) Sprague in F.T.A. 6(1): 312 (1910); F.D.O.-A. 2: 166 (1932), pro parte; Balle in F.C.B. 1: 319 (1948); T.T.C.L.: 282 (1949)

NOTE. Despite the general similarity to *E. dregei* these populations seem sufficiently distinctive, uniform and geographically discrete to warrant specific rank. The differences are least perhaps with the forms of *E. dregei* extending up the western side of the Rift Valley from Elgon to N. Ethiopia as pointed out by Sprague (1910) when he ranked it as a variety. The populations of *E. dregei* in the peripheral *Brachystegia* woodlands are notably polymorphic and may approach the robustness of *E. taborense* at least in their leaves, but obvious hybrids have not been noted yet. The species flowers in the dry season, May–October, later than *E. dregei* where the ranges overlap.

4. **E. lanatum** *Wiens & Polh.*, Mistletoes Afr.: 241 (1998). Type: Zambia, Mbala District, Kawimbe, *Richards* 6982 (K!, holo.)

Twigs tomentose with cream or buff stellate, dendritic and longer wispy subsimple hairs, glabrescent. Leaves opposite, soon clustered on short shoots; petiole 4–5 mm. long; lamina oblong-elliptic to broadly elliptic, 3.5–9 cm. long, 1.5–6 cm. wide, broadly cuneate to shortly cordate at the base, felty stellate-tomentose on both surfaces. Heads terminal from a tuft of normally 2 pairs of leaves or clustered on older leafless nodes, 4-flowered (occasionally 1 below the others); peduncle 1–4 cm. long; bract ovate, acuminate at the apex, concave, 5–8(–10) mm. long. Receptacle 1.5–2 mm. long, long-hairy; calyx 2.5–3 mm. long, silky villous. Corolla 4.5–5.5 cm. long, pale green or cream with the lobes tipped green, silky villous, densely so above the basal swelling; basal swelling 6–7 mm. long; constriction 4–5 mm. long; funnel-shaped upper part 7–9 mm. long. Berry not seen.

TANZANIA. Ufipa District: Lake Kwela, 4 Nov. 1956, *Richards* 6864!
DISTR. **T** 4; N. Zambia
HAB. *Brachystegia* woodland, recorded on *Erythrina*; 1800 m.

5. **E. schmitzii** *Wiens & Polh.*, Mistletoes Afr.: 241 (1998). Type: Zaire, Shaba, Lubumbashi Golf Course, *Schmitz* 863 (BR!, holo.)

Stems to 1 m. long; twigs tomentose with stellate and short dendritic hairs, soon glabrescent. Leaves opposite and also crowded on short shoots; petiole 5–10 mm. long; lamina narrowly oblong-elliptic, 3.5–7 cm. long, 0.8–1.5 cm. wide, obtuse to rounded at the apex, cuneate at the base, stellate-tomentose at first, soon glabrescent, with 8–10 pairs of obscure lateral nerves. Heads terminal from a tuft of leaves, (2–)4-flowered; peduncle 1–1.5 cm. long, tomentose; bract ovate-triangular, sometimes acuminate into a short incurved limb, 3–8 mm. long. Receptacle 1.5 mm. long, long-hairy; calyx 2–2.5 mm. long, silky-pubescent, ± glabrescent. Corolla 4–5 cm. long, pale green at the base, orange on the tube above, with lobes pale greeny yellow, silky villous, the hairs especially long and somewhat evanescent on the basal swelling; basal swelling 5–6 mm. long; constriction 3 mm. long; funnel-shaped upper part 5–8 mm. long. Berry 10 mm. long, 7 mm. in diameter, pilose and rugose to squamate around the middle.

TANZANIA. Mbeya District: Pungaluma, 17 Dec. 1989, *Congdon* 230!; Iringa District: Iheme, 25
  Feb. 1962. *Polhill & Paulo* 1606! & 0.7 km. S. of Tanzam highway on W. loop-road to Dabaga,
  29 Jan. 1986, *Wiens & Frank* 6593!
DISTR. **T** 7; Zaire (Shaba)
HAB. Higher rainfall *Brachystegia* woodland and edges of montane forest, generally on
  *Brachystegia* or *Julbernardia*; 1800–1900 m.

6. **E. rotundifolium** *Wiens & Polh.*, Mistletoes Afr.: 242 (1998). Type: Tanzania, Kilwa District, Selous Game Reserve, Nahomba valley, *Vollesen* in *M.R.C.* 4794 (K!, holo., C!, DAR!, EA!, WAG!, iso.)

Twigs tomentose with somewhat fulvous stellate and shortly dendritic hairs, soon glabrescent. Leaves opposite and clustered on short shoots, sessile or nearly so, glaucous, subcircular, 1.5–5 cm. long, 1.5–4 cm. wide, cordate at the base, early glabrescent, with 4–5 pairs of lateral nerves. Heads terminal on a short shoot of normally 1 pair of leaves, sometimes also axillary, (2–)4-flowered; peduncle 1.5–2 cm. long, glabrous; bract ovate, slightly acuminate at the apex, concave, 4–5 mm. long, subglabrous apart from margins. Receptacle 1.5 mm. long, silky villous; calyx 3–4 mm. long, subglabrous except for a fringe of small brown hairs. Corolla 3.5–4 cm. long, orange, with the lobes and sometimes the basal part of tube green, silky villous in bud but glabrescent on basal swelling and to varying degrees on the lobes; basal swelling 5–7 mm. long; constriction 3–4 mm. long; distal part of tube ± 5 mm. long. Berry not seen.

TANZANIA. Kilwa District: Selous Game Reserve, Nahomba valley, 3 Dec. 1977, *Vollesen* in *M.R.C.*
  4794!
DISTR. **T** 8; known only from the type gathering
HAB. *Brachystegia* woodland, on *Brachystegia*; 500 m.

SYN. *Erianthemum sp. nov.* sensu Vollesen in Opera Bot. 59: 64 (1980)

7. **E. nyikense** (*Sprague*) *Danser* in Verh. K. Akad. Wet., sect. 2, 29(6): 54 (1933); Polh. & Wiens, Mistletoes Afr.: 242, photo. 108 (1998). Type: Malawi, Nyika Plateau, *McClounie* 111 (K!, holo., B!, iso.)

Stems spreading to 1 m. or so; twigs tomentose, with stellate and ± dendritic somewhat fulvous hairs, glabrescent. Leaves opposite to subopposite, soon clustered on short shoots; petiole 0–9 mm. long; lamina elliptic or oblong-elliptic to ovate, (1–)2–5 cm. long, (0.8–)1.2–4 cm. wide, mostly rounded to cordate at the base, in very young leaves sometimes cuneate, densely stellate-pubescent to -tomentose on both surfaces, somewhat glabrescent to varying degrees, with ± 4 obscure pairs of lateral nerves. Heads mostly terminal from a tuft of 1–2 pairs of leaves, but also axillary on younger branches, (2–)4-flowered; peduncle 3–25(–30) mm. long; bract with a well-developed leafy elliptic-obovate to elliptic-circular limb, (10–)15–25 mm. long, (5–)8–20 mm. wide. Receptacle 1 mm. long, long-hairy; calyx 2–3 mm. long, silky villous. Corolla 3.5–5 cm. long, greenish, tube flushed yellowish to orange, densely silky villous except on lower part of basal swelling; basal swelling 4–5 mm. long; constriction 3–4 mm. long; funnel-shaped upper tube 7–9 mm. long. Berry with calyx 10 mm. long, pinkish red, pilose.

TANZANIA. Rungwe District: Kaporogwe, Makete Waterfall, 30 Dec. 1987, *Lovett & Thomas* 2866! & Kyimbila, 22 Nov. 1910, *Stolz* 421!
DISTR. **T** 7; Zambia, Malawi, Mozambique and Zimbabwe
HAB. Montane forest edges and along rivers; 1000–1350 m.

SYN. *Loranthus nyikensis* Sprague in F.T.A. 6(1): 315 (1910) & in K.B. 1911: 140 (1911); F.D.O.-A. 2: 168 (1932); T.T.C.L.: 283 (1949)

8. **E. lindense** (*Sprague*) *Danser* in Verh. K. Akad. Wet., sect. 2, 29(6): 54 (1933); Polh. & Wiens, Mistletoes Afr.: 243 (1998). Type: Tanzania, Lindi, *Busse* 3005 (B!, holo., BR!, iso., K fragment & drawing!)

Stems to 1 m. or so; twigs tomentose with stellate and some short dendritic somewhat fulvous hairs, glabrescent. Leaves opposite, becoming clustered at older nodes; petiole 3–5 mm. long; lamina ovate to oblong-elliptic, 1.5–7 cm. long, 0.8–5 cm. wide, mostly rounded to cordate at the base, in very young leaves sometimes cuneate, stellate-pubescent on both surfaces, variably glabrescent, with ± 4 obscure pairs of lateral nerves. Heads 1(–several) in axils or terminal from a tuft of 1–2 pairs of leaves, (2–)4-flowered; peduncle (1.5–)2–4 cm. long; bract with a small to well-developed leafy limb, (7–)10–20 mm. long, (4–)5–14 mm. wide. Receptacle 1 mm. long, long-hairy; calyx 3–4.5 mm. long, glabrous except for fringe of dendritic or mostly simple hairs around the top. Corolla 3.5–5 cm. long, pale green, the tube flushed orange to pink, densely silky villous except on the lower part of the basal swelling; basal swelling 5–6 mm. long; constriction 3–4 mm. long; funnel-shaped upper tube 5–8 mm. long. Berry deep pink, subglabrous.

TANZANIA. Songea District: 1.5 km. E. of Songea, 15 Jan. 1956, *Milne-Redhead & Taylor* 8342!; Masasi District: Masasi Hill, 8 km. NE. of Masasi, 18 Mar. 1991, *Bidgood, Abdallah & Vollesen* 2067!; Lindi District: Rondo Forest Reserve, W. side, 22 Nov. 1966, *Gillett* 18001!
DISTR. **T** 8; N. Mozambique
HAB. *Brachystegia* woodland and coastal associations; 300–1100 m.

SYN. *Loranthus lindensis* Sprague in F.T.A. 6(1): 316 (1910) & in K.B. 1911: 141 (1911); F.D.O.-A. 2: 168 (1932); T.T.C.L.: 282 (1949)
    *L. dregei* Eckl. & Zeyh. var. *longipes* Sprague in F.T.A. 6(1): 314 (1910) & in K.B. 1911: 140 (1911); F.D.O.-A. 2: 167 (1932); T.T.C.L.: 282 (1949). Type: Tanzania, Songea District, Kwa Kihingi, *Busse* 764 (B!, holo., EA!, iso., K!, fragment)
    [*Erianthemum ngamicum* sensu Vollesen in Opera Bot. 59: 63 (1980), *non* (Sprague) Danser]

NOTE. Closely related to *E. nyikense* and replacing it north-eastwards. It has been confused with *E. dregei*, but the combination of long peduncles, foliaceous bracts and subglabrous calyx is distinctive, apart from the clustered leaves (these may not be apparent on gatherings of young shoots only).

9. **E. alveatum** (*Sprague*) *Danser* in Verh. K. Akad. Wet., sect. 2, 29(6): 53 (1933); Polh. & Wiens, Mistletoes Afr.: 243 (1998). Type: Tanzania, Tanga District, Doda, *Holst* 2946 (B!, holo., K!, LISC!, iso.)

Small shrub; twigs tomentose with stellate and short dendritic slightly fulvous hairs, soon glabrescent. Leaves opposite to alternate, soon clustered on short shoots; petiole 3–5 mm. long; lamina oblong-elliptic to oblong-oblanceolate, 1.5–4 cm. long, 0.5–2 cm. wide, cuneate to shortly rounded at the base, stellate-pubescent, early glabrescent, with 4–5 pairs of lateral nerves. Heads terminal on short shoots (commonly composed of 2 pairs of leaves) and in axils on younger long shoots, (2–)4-flowered, peduncle 1–2.5 cm. long, sparsely hairy, bract with a leafy limb, 8–12 mm. long, 4–8 mm. wide. Receptacle 1.5 mm. long, long-hairy; calyx 2.5–3 mm. long, silky villous. Corolla 3–4 cm. long, green and apricot, densely silky villous except for the lower part of the basal swelling; basal swelling 5 mm. long; constriction 3 mm. long; funnel-shaped upper tube 7–8 mm. long. Berry not seen.

KENYA. Kwale District: 13 km. N. of Lungalunga, 18 Jan. 1972, *Wiens* 4532! & 61 km. from Mombasa on Nairobi road, 20 Jan. 1961, *Greenway* 9791!; Kilifi District: Mangea Hill, 9 July 1987, *S.A. Robertson & Luke* 4925A!
TANZANIA. Tanga District: Doda, June 1893, *Holst* 2946!
DISTR. **K** 7; **T** 3; known only from coastal belt between Malindi and Tanga
HAB. Coastal bushland, woodland and wooded grassland, recorded on *Brachystegia* and *Terminalia*; 0–300 m.

SYN. *Loranthus alveatus* Sprague in F.T.A. 6(1): 315 (1910) & in K.B. 1911: 141 (1911); F.D.O.-A. 2: 168 (1932); T.T.C.L.: 281 (1949)

10. **E. commiphorae** (*Engl.*) *Danser* in Verh. K. Akad. Wet., sect. 2, 29(6): 53 (1933); Polh. & Wiens in U.K.W.F., ed. 2: 157 (1994) & Mistletoes Afr.: 244, photo. 109A–B, fig. 26D (1998). Type: Tanzania, Iringa District, Lukose R., *Goetze* 477 (B!, holo., BR!, K!, iso.)

Stems well branched, to 1 m. long; twigs tomentose with stellate and sometimes some short dendritic hairs, glabrescent. Leaves opposite or subopposite, soon clustered on short shoots; petiole 1–2 mm. long; lamina ovate-lanceolate, oblong-lanceolate or oblong-oblanceolate, 1.7–3.5 cm. long, 0.4–1.5 cm. wide, narrowed to the base, stellate-pubescent on both surfaces, obscurely nerved. Heads terminal on short shoots, (2–)4-flowered; peduncle 1.5–2.5 cm. long; bract leafy, oblong-elliptic, 5–20 mm. long, 4–6 mm. wide. Receptacle 1.5–2 mm. long, white hairy; calyx 1–1.5 mm. long, tomentose at least towards the top. Corolla 4–5 cm. long, greenish yellow to pale orange, sometimes marked pink or red, the basal part sometimes green, the lobes yellow to slightly orange, stellate tomentose with some longer little branched hairs on the tube, the basal swelling nearly glabrous at least below, the lobes sparsely stellate-pubescent and glabrescent; basal swelling 4–5 mm. long; constriction 1–2 mm. long, hairy inside; upper dilated part of tube 6–8 mm. long. Berry blue, 10–12 mm. long, 7 mm. in diameter, glabrescent; seed red. Fig. 17/5–9 (p. 97).

KENYA. Machakos District: Tsavo National Park, Chyulu Hills, 17 km. NW. of Chyulu Gate, 23 Jan. 1972, *Wiens* 4542! & Kibwezi, Mar. 1922, *Dummer* 5030!; Masai District: 19 km. along old Kajiado road, 6 Nov. 1956, *Bally* 10456!
TANZANIA. Masai District: Olduvai Gorge, 16 Jan. 1990, *Chuwa* 3090!; Kondoa District: between Kolo and Chungai, 24 km. N. of Kondoa, 13 Jan. 1962, *Polhill & Paulo* 1159!; Iringa District: Kibebe Farm, 30 Dec. 1995, *Congdon* 463!
DISTR. **K** 4, 6, 7; **T** 2, 5–7; not known elsewhere
HAB. Deciduous bushland, on a variety of hosts, including *Acacia, Commiphora, Euphorbia* and *Lannea*; 700–1800 m.

SYN. *Loranthus commiphorae* Engl. in E.J. 28: 380 (1900); Sprague in F.T.A. 6(1): 308 (1910); F.D.O.-A. 2: 165 (1932), pro majore parte; T.T.C.L.: 281 (1949)

11. **E. occultum** (*Sprague*) *Danser* in Verh. K. Akad. Wet., sect. 2, 29(6): 54 (1933); Polh. & Wiens, Mistletoes Afr.: 244 (1998). Type: Tanzania, Tanga District, Udigo, between Umba and Nyika, *Busse* 1128 (K!, holo., B!, iso.)

Stems spreading to 50 cm. or so; twigs white stellate-tomentose, soon glabrescent. Leaves opposite or subopposite, soon clustered on short shoots, often sparse at flowering time; petiole 2–5 mm. long; lamina oblanceolate to elliptic-obovate or obovate, 1–3 cm. long, 0.5–1.8 cm. wide, cuneate at the base, stellate-pubescent, soon glabrescent, with 3–4 pairs of lateral nerves. Heads terminal on short shoots, 2–4-flowered; peduncle 5–20 mm. long; bract leafy, oblong-oblanceolate, 10–20 mm. long, 4–10 mm. wide, glabrous except near the base. Receptacle 1–1.5 mm. long, long white hairy; calyx 0.5–1.5 mm. long, similarly hairy. Corolla 3–4.5 cm. long, yellow flushed orange to red on the tube, subglabrous or early glabrescent except for scattered stellate or wispy hairs; basal swelling 3–5 mm. long; constriction 1–2 mm. long, hairy inside; upper dilated part of tube 6–8 mm. long. Berry not seen.

KENYA. Northern Frontier Province: 20 km. S. of Moyale, 15 Oct. 1952, *Gillett* 14049!; Machakos District: 5 km. E. of Mtito Andei on Mombasa road, 9 Nov. 1958, *Greenway* 9542!; Teita District: Buchuma, 17 Oct. 1982, *R. & D. Polhill* 4864!
TANZANIA. Lushoto District: 16 km. SE. of Mkomazi on main road to Mombo, 30 Dec. 1985, *D. & C. Wiens* 6525!; Tanga District: Udigo, between Umba and Nyika, Oct. 1900, *Busse* 1128!; Iringa District: Iringa–Dodoma road, Isimani, 31 Jan. 1988, *Lovett* 2992!
DISTR. **K** 1, 4, 7; **T** 3, 7; not known elsewhere
HAB. Deciduous bushland, recorded only on *Commiphora*; 150–1100 m.

SYN. *Loranthus occultus* Sprague in F.T.A. 6(1): 308 (1910) & in K.B. 1911: 137 (1911); T.T.C.L.: 283 (1949)
    [*L. commiphorae* sensu F.D.O.-A. 2: 165 (1932), pro parte, quoad specim. *Busse* 1128, *non* Engl.]


12. **E. aethiopicum** *Wiens & Polh.* in Nordic Journ. Bot. 5: 223 (1985); M.G. Gilbert in Fl. Ethiopia 3: 366, t. 114.7/1–5 (1990); Polh. & Wiens, Mistletoes Afr.: 245, photo. 110 (1998). Type: Ethiopia, Ogaden, 24 km. SW. of Werder [Wardere], *Ellis* 142 (K!, holo.)

Twigs tomentose with white stellate and short dendritic hairs, glabrescent. Leaves opposite or subopposite, soon clustered on short shoots, often sparse at flowering time; petiole (1–)2–3 mm. long; lamina elliptic to oblanceolate, 5–25 mm. long, 1.5–8 mm. wide, cuneate at the base, stellate-pubescent on both surfaces, gradually glabrescent, with 4–6 sometimes obscure pairs of lateral nerves. Heads terminal on short shoots or on older leafless nodes, (2–)4-flowered; peduncle 5–15 mm. long; bract broadly ovate or obovate, concave, 3–6 mm. long, sometimes with a short leafy limb and then up to 10 mm. long, stellate tomentose. Receptacle (0.5–)1–1.2 mm. long, white tomentose; calyx 0.7–1 mm. long, similarly hairy. Corolla 3–4.5 cm. long, yellow or yellow-green, the tube becoming flushed orange or red, the basal swelling villous, the rest densely covered with a mixture of stellate, dendritic and relatively few long simple hairs; basal swelling 3–4 mm. long; constriction ± 1 mm. long, hairy inside; upper part of tube 7–9 mm. long. Berry blue-green, 10 mm. long including calyx, sparsely pubescent; seed red.

KENYA. Northern Frontier Province: Dandu, 25 Mar. 1952, *Gillett* 12636! & Ramu–Banissa road, 59 km. from the turning to Banissa, 4 May 1978, *M.G. Gilbert & Thulin* 1447! & 48 km. on the Ramu–El Wak road, 9–10 May 1978, *M.G. Gilbert & Thulin* 1601!
DISTR. **K** 1; Ethiopia and Somalia
HAB. Deciduous bushland, generally on *Commiphora*, but also recorded on *Acacia*, *Euphorbia* and *Ipomoea*; 650–850 m.

## 16. PHRAGMANTHERA

Tiegh. in Bull. Soc. Bot. Fr. 42: 261 (1895); Balle in Webbia 11: 583 (1955); Polh. & Wiens, Mistletoes Afr.: 246 (1998)

Shrubs, often large and pendent, with a single haustorial attachment; branchlets slightly compressed; youngest parts, at least, always with scales, stellate and verticillately branched dendritic hairs. Leaves opposite, subopposite or rarely ternate, generally petiolate, usually somewhat coriaceous, penninerved. Flowers in sessile to shortly pedunculate umbels or 2–several in the axils; bract unilateral, often gibbous, sometimes enlarged and foliaceous. Calyx annular to shortly tubular, generally shorter than receptacle, subentire to shortly lobed. Corolla 5-lobed, with tube longer than lobes, usually yellow or orange with red markings or more generally reddish, but colours often muted by indumentum; apical swelling of bud fusiform to globular, sometimes ribbed or narrowly winged; basal swelling variably developed; tube split unilaterally, the V-slit extending little more or less than halfway, papillate along sutures and edges of filament-lines to varying degrees; lobes erect or reflexed, narrow below, the upper part expanded to varying degrees and often hardened inside. Stamens attached near the base of the lobes, linear, inrolled at anthesis, sometimes with a tooth in front of the anther; anther oblong to linear-oblong, with 4 thecae transversely divided into several–many chambers. Style slender, 5-angled or narrowly 5-winged, often expanded opposite the filaments and narrowed opposite anthers, sometimes papillate or hairy; stigma globular to obovoid or conical-ovoid. Berry reddish or often blue to blue-green, ellipsoid or obovoid, generally with persistent calyx.

34 species, well distributed in tropical Africa and Arabia, the main diversity around the edges of the Guineo-Congolian forests, especially from Nigeria to Angola. The species are often very common locally and liable to become pests of plantation crops.

*Phragmanthera* closely mimics *Agelanthus* in its flower structure, and both genera were included in *Tapinanthus* by Danser in Verh. K. Akad. Wet., sect. 2, 29(6): 22, 107 (1933). It is easily distinguished from both those genera by the indumentum of scales, stellate and dendritic hairs, by the unilateral bract and locellate anthers. It appears most closely related to *Oedina* and the Asiatic genus *Dendrophthoë*, differing from both by the more specialised pollination mechanisms.

The flowers open by vents developing between the lower part of the corolla-lobes, revealing the contrasting colour of the filaments as a honey-guide. Penetration of the pollinator's beak causes the characteristic V-shaped slit on that side, the sprung filaments incurve to scatter pollen and the stigma is often pulled towards the bird's head by the curling stamens. After anthesis the conspicuousness of the inflorescence is often enhanced by intensive colouring of the corolla-tube inside. The most salient characters for sectional and species recognition tend to be related to the pollination syndrome, notably the shape, length and elasticity of the corolla-lobes, the development of a tooth in front of the anther, and details of the shape and ornamentation of the style. The density, distribution and colour of the hairs on the corolla affect the general appearance of the flowers and often show precise differences between related species and subspecies.

Most of the species in East Africa belong to sect. *Rufescentes* (Engl.) Polh. & Wiens (syn. *Loranthus* sect. *Dendrophthoë* group *Rufescentes* Engl.; *L.* sect. *Rufescentes* (Engl.) Sprague), but *P. eminii* and *P. proteicola* belong to the distinctive section *Eubracteatae* (Engl.) Polh. & Wiens (syn. *Loranthus* subgen. *Tapinanthus* group *Eubracteati* Engl.; *L.* sect. *Eubracteati* (Engl.) Sprague).

Certain species of *Agelanthus* from southern Tanzania with brownish irregularly branched hairs, notably *A. songeensis* and *A. uhehensis*, are liable to be confused with species of *Phragmanthera* if not examined carefully.

1. Corolla with lobes erect and generally with a distinct basal swelling, the ground-colour of the tube (excluding hairs) yellow to brown or red; filaments not toothed (sect. *Rufescentes*, fig. 18) · · · · · · · · · · · · · · · · · · · · · · · · · · · · 2
   Corolla with lobes reflexed and basal swelling slight, ground-colour of tube green (sometimes flushing red); filaments with a tooth in front of the anther (sect. *Eubracteatae*, fig. 19) · · · · · · · · · · · · · · · · · · · · · · · · · · · 7
2. Leaves becoming clustered on short shoots, narrow, 4.5–15 times as long as broad; style generally distinctly lobed opposite base of filaments · · · · · · · · · · · · · · · · · · · · · · · · 5. *P. dschallensis*
   Leaves not crowded (sometimes ternate), 1.1–4.5 times as long as broad; style not lobed · · · · · · · · · · · · · · · · · · · · · · · · · · 3
3. Hardened inner surface of corolla-lobes extended well below the expanded part, thus distinctly spathulate above the narrow infolded claw; corolla yellow to bright orange, sometimes flushing red on tube and tip of lobes, subglabrous to thinly puberulous in East Africa · · · · · · · · · · · · · · · · · · · 4
   Hardened inner surface of corolla-lobes restricted to expanded part and often ± 1 mm. below (corresponding to underside of apical swelling in bud), cuneate to the long narrow infolded claw in mature open flowers; corolla reddish (part sometimes muted yellow), flushing deeper and often conspicuously hairy (except *P. usuiensis* subsp. *sigensis*) · · · · · · · · · · · · · · · · · · · · · · · · · · · · · · · · · · · · · · · 5
4. Leaves not markedly discolorous; corolla-lobes 12–15 mm. long, slightly winged · · · · · · · · · · · · · · · · · · · · · · · · · · 1. *P. brieyi*
   Leaves markedly discolorous, the lamina green, glossy and early glabrescent above, persistently reddish lepidote-tomentellous beneath; corolla-lobes 9–12 mm. long, not winged · · · · · · · · · · · · · · · · · · · · · · · · · · · · · · · · · · 6. *P. polycrypta*
5. Corolla-lobes 12–16 mm. long · · · · · · · · · · · · · · · · · · · · · 4. *P. regularis*
   Corolla-lobes 8–12 mm. long · · · · · · · · · · · · · · · · · · · · · · · · · · · 6
6. Corolla tomentellous to subglabrous with hairs mostly less than 0.5 mm. long, glabrescent in part; branchlets with tomentum persisting only on first 1–few internodes: style generally only slightly broadened upwards to shoulders · · · 2. *P. usuiensis*
   Corolla tomentose with hairs 1–3 mm. long and rather persistent overall; branchlets subpersistently hairy; style abruptly expanded to twice its diameter opposite base of filaments, then tapered slightly to shoulders · · · · · · · · · · 3. *P. cornetii*
7. Pedicels 2–4 mm. long; bracts mostly small, 3–6 mm. long, occasionally some foliaceous; anthers 2–3 mm. long; branchlet-hairs 0.5–1 mm. long · · · · · · · · · · · · · · · · · · 7. *P. eminii*
   Pedicels 0.5–1 mm. long; bracts spathulate to foliaceous, mostly 1–1.5 cm. long; anthers 3–4 mm. long; branchlet-hairs 1–4 mm. long · · · · · · · · · · · · · · · · · · · · · · · · · · · · 8. *P. proteicola*

1. **P. brieyi** (*De Wild.*) *Polh. & Wiens* in Lebrun & Stork, Énum. Pl. Fl. Afr. Trop. 2: 170 (1992) & Mistletoes Afr.: 255 (1998). Type: Zaire, Mayumbe, Ganda Sundi, *de Briey* 1025 (BR!, holo.)

Much-branched shrub; branchlets with youngest parts densely pale lepidote and very youngest with small yellow-brown dendritic hairs less than 0.3 mm. long, soon glabrescent. Petiole 5–16 mm. long; lamina coriaceous, oblong-lanceolate to ovate-oblong or ovate, 3–12 cm. long, 1.5–8 cm. wide, obtuse at the apex, rounded to cuneate at the base, initially covered with pale to yellow-brown stellate and very short dendritic hairs, soon glabrescent apart from sometimes a scatter of pale stellate hairs especially beneath, with 3–8 pairs of lateral nerves. Umbels axillary and at older nodes, 2–6-flowered; peduncle 0–3 mm. long; pedicels 2–3 mm. long, pale lepidote-tomentellous, glabrescent in patches; bract ovate-elliptic to round, 1–2 mm. long, glabrescent. Receptacle urceolate, 2–3 mm. long, pale lepidote-tomentellous; calyx 0.2–0.8 mm. long, slightly lobed, glabrescent. Corolla 4–4.5 cm. long, orange-yellow to bright orange, puberulous with minute stellate hairs, glabrescent especially distally; apical swelling of bud ellipsoid, 4–5 mm. long, 3 mm. in diameter, with wings to 0.5 mm. wide; basal swelling slight, 7 mm. long, 3.5 mm. in diameter, lobes erect, 12–15 mm. long, upper half spathulate and hardened inside; tips 1 mm. wide, slightly winged. Anthers 1.5–3 mm. long, with 5–10 cells in each row. Style almost isodiametric, sometimes puberulous opposite the corolla-tube, with a slight neck 2.5–3.5 mm. long below the conical-globose stigma. Berry not seen.

TANZANIA. Bukoba District: Rusinga Forest Reserve, 24 Nov. 1994, *Congdon* 387!
DISTR. **T** 1; Zaire and N. Angola (Mayombe)
HAB. Swamp forest; 1200 m.

SYN. *Loranthus brieyi* De Wild. in B.J.B.B. 4: 410 (1914); Balle in F.C.B. 1: 359 (1948)
    *Tapinanthus brieyi* (De Wild.) Danser in Verh. K. Akad. Wet., sect. 2, 29(6): 109 (1933)

NOTE. *Phragmanthera brieyi* is otherwise known only from a few specimens collected along the middle and lower reaches of the Congo R. The single specimen from Tanzania, collected fairly recently and not included in our account for the Mistletoes Afr. (1998), agrees in essential characters with this distinctive species, but differs in some respects, the hairs less obviously white, and the styles slightly more swollen in the middle part and lacking hairs seen in western material.

2. **P. usuiensis** (*Oliv.*) *M.G. Gilbert* in Nordic. Journ. Bot. 5: 224 (1985) & in Fl. Ethiopia 3: 368 (1990); Polh. & Wiens in U.K.W.F., ed. 2: 157, t. 57 (1994) & Mistletoes Afr.: 259 (1998). Type: Tanzania, Biharamulo District, Usui, *Grant* (K!, holo.)

Branches spreading to pendent, 1–2 m. long; branchlets tomentose with short white to brown dendritic hairs on youngest internodes, soon glabrescent with only grey stellate hairs and scales, then glabrous. Petiole 1–3 cm. long; lamina dark green, paler beneath, coriaceous, oblong-lanceolate or oblong-elliptic to elliptic or ovate, 5–20 cm. long, 2–11 cm. wide, obtuse to rounded at the apex, cuneate to rounded or cordate at the base, soon glabrescent, with 6–8 pairs of inconspicuous lateral nerves. Umbels confluent in older axils and mostly around nodes below the current leaves, 2–6-flowered; peduncle 0–2(–3) mm. long; pedicels 1–3 mm. long; bract ovate or ovate-lanceolate, ± acuminate at the apex, 2–4 mm. long, hairy to varying degrees. Receptacle cylindrical, 1.2–2.5 mm. long, tomentellous; calyx 0.5–1.5 mm. long, usually glabrescent, ciliate. Corolla 3–5 cm. long, yellow to orange, reddening on basal and especially apical swelling of bud, reddening on tube inside, brownish pubescent to tomentellous with stellate and ± short dendritic hairs in bud, but glabrescent to varying degrees at least on the head and basal swelling, sometimes nearly glabrous overall; apical swelling of bud ovoid-ellipsoid to nearly globular, slightly apiculate and 5-ribbed, 3–3.5(–4) mm. long, 2–2.5(–3) mm. in diameter; basal swelling 3–6 mm. long, 2–3 mm. in diameter, with the tube generally distinctly narrowed for several mm. above; lobes erect, 8–12 mm. long; tips elliptic, apiculate, 3–4 mm. long, 1.5 mm. wide, hardened inside. Filaments red; anthers 1–2 mm. long, with 3–6 cells in each row. Style red above, slightly broadened upwards to shoulder; neck 1–2 mm. long. Berry blue-green, broadly ellipsoid, 10 mm. long, 6–8 mm. in diameter, glabrescent; seed bright orange-red.

subsp. **usuiensis**; Polh. & Wiens, Mistletoes Afr.: 260, photo. 113A–B, fig. 27B (1998)

Corolla persistently tomentellous except on head and later basal swelling, or glabrescent to varying degrees but with at least some dendritic hairs on narrow part of the tube. Calyx generally half as long as receptacle at anthesis. Twigs with indumentum (at least stellate hairs and scales) often persistent for several internodes. Fig. 18/1–11.

UGANDA. Toro District: Ruwenzori, Nyinabitaba [Nyabitaba], Jan. 1951, *Osmaston* 3737!; Teso District: Serere, Jan. 1932, *Chandler* 394!; Mengo District: Kawanda Research Station, 25 Mar. 1971, *McNutt* 126!
KENYA. Nakuru District: Njoro R. valley, 21 Jan. 1985, *Kirkup* 14!; N. Kavirondo District: Kakamega Forest, Rondo, 6 Oct. 1982, *R. & D. Polhill* 4859!; Masai District: 2.5 km. W. of Ngore Ngore and Keekorok road junction, 20 Dec. 1985, *D. & C. Wiens* 6510!
TANZANIA. Musoma District: crest of Lobo range, 18 May 1961, *Greenway, Myles Turner & Owen* 10310!; Mpanda District: NW. slopes of Musenabantu, 12 Aug. 1959, *Harley* 9305!; Lindi District: Rondo Forest Reserve, 8 km. NW. of Forest Station, 17 Feb. 1991, *Bidgood, Abdallah & Vollesen* 1605!
DISTR. U 1–4; K 2–6; T 1, 2, 4, 7, 8; widespread from Cameroon to Ethiopia and south to S. Zaire, Angola and W. Zambia, approaching the range of subsp. *sigensis* only in S. Tanzania
HAB. Forests and adjacent woodland, on a wide variety of hosts including many cultivated trees; 850–3000 m.

SYN. *Loranthus usuiensis* Oliv. in Trans. Linn. Soc., Bot., 29: 80, t. 44 (1873); P.O.A. C: 165 (1895); Sprague in F.T.A. 6(1): 288 (1910); F.D.O.-A. 2: 162 (1932), excl. var. *longipilosus*; T.T.C.L.: 288 (1949), excl. var.; F.P.S. 2: 292 (1952)
    *L. bukobensis* Engl. in E.J. 20: 102 (1894) & P.O.A. C: 166 (1895); Engl. & K. Krause in Z.A.E. 2: 195 (1911). Lectotype, chosen by Polh. & Wiens (1998): Tanzania, Bukoba, *Stuhlmann* 1092 (B!, lecto.)
    *L. albizziae* De Wild., Miss. E. Laurent: 74 (1905); Sprague in F.T.A. 6(1): 288 (1910), excl. var. *rogersii*; Balle in F.C.B. 1: 364, t. 36 (1948); F.P.N.A. 1: 101 (1948). Type: Zaire, Basoko, *E. & M. Laurent* (BR!, holo.)
    [*L. regularis* sensu Engl. & K. Krause in Z.A.E. 2: 196 (1911); Balle in F.C.B. 1: 363 (1948), *non* Schweinf.]
    *L. kisaguka* Engl. & K. Krause in E.J. 51: 460 (1914); F.D.O.-A. 2: 162 (1932); T.T.C.L.: 288 (1949). Types: Tanzania, Rungwe District, Mt. Rungwe, *Stolz* 389 & Rutangania, *Stolz* 420 (both B!, syn., K!, WAG!, isosyn.)
    *L. usuiensis* Oliv. var. *maitlandii* Sprague in K.B. 1916: 179 (1916). Type: Uganda, Lake Districts, *Maitland* 119 (K!, holo.)
    *L. rigidilobus* K. Krause in N.B.G.B. 8: 497, fig. 2E–K (1923); Chiov., Racc. Bot. Miss. Consol. Kenya: 109 (1935). Type: Kenya, Meru, *R.E. & T.C.E. Fries* 1711 (B!, holo., K!, iso.)
    *Tapinanthus albizziae* (De Wild.) Danser in Verh. K. Akad. Wet., sect. 2, 29(6): 107 (1933)
    *T. kisaguka* (Engl. & K. Krause) Danser in Verh. K. Akad. Wet., sect. 2, 29(6): 114 (1933)
    *T. rigidilobus* (K. Krause) Danser in Verh. K. Akad. Wet., sect. 2, 29(6): 118 (1933)
    *T. usuiensis* (Oliv.) Danser in Verh. K. Akad. Wet., sect. 2, 29(6): 121 (1933)
    [*Loranthus rufescens* sensu Balle in F.C.B. 1: 367 (1948); F.P.N.A. 1: 101 (1948); U.K.W.F.: 330, fig. on 329 (1974), *non* DC.]
    *Phragmanthera albizziae* (De Wild.) Balle in B.J.B.B. 37: 457 (1967)
    *P. rufescens* (DC.) Balle subsp. *usuiensis* (Oliv.) Balle in Fl. Rwanda 1: 185, fig. 38/3 (1978); Troupin, Fl. Pl. Lign. Rwanda: 375, fig. 132/3 (1982), *nom. invalid., comb. non rite publ.*
    [*P. regularis* sensu Blundell, Wild Fl. E. Afr.: 130 (1987), *non* (Schweinf.) M.G. Gilbert]
    *P. regularis* (Schweinf.) M.G. Gilbert var. *usuiensis* (Oliv.) Blundell, Wild Fl. E. Afr.: 130, t. 470 (1987), as '*usuriensis*', *nom. invalid., comb. non rite publ.*

subsp. **sigensis** (*Engl.*) *Polh. & Wiens* in Lebrun & Stork, Enum Pl. Fl. Afr. Trop. 2: 172 (1992) & Mistletoes Afr.: 260 (1998). Type: Tanzania, Tanga District, E. Usambara Mts., Derema, *Holst* 2230 (B!, holo., K!, iso.)

Corolla glabrescent early, generally with only appressed subsimple hairs (occasionally very short dendritic hairs). Calyx generally two-thirds as long as receptacle at anthesis. Twigs (except southernmost populations) generally glabrescent by second internode.

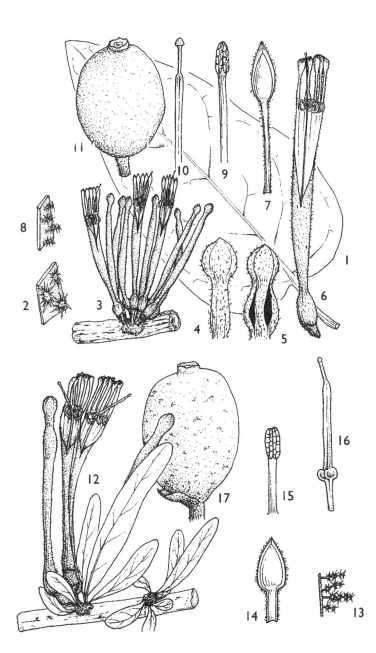

FIG. 18. *PHRAGMANTHERA USUIENSIS* subsp. *USUIENSIS* — **1**, leaf, × 1, **2**, scales from surface of twig, × 20; **3**, flowering node, × 1; **4**, tip of young flower bud, × 3; **5**, same at maturity, with vents open, × 3; **6**, flower, × 2; **7**, detail of corolla-lobe, × 4; **8**, surface of corolla showing hairs, × 20; **9**, stamen, × 4; **10**, style-tip, × 4; **11**, fruit, × 2. *PHRAGMANTHERA DSCHALLENSIS* — **12**, flowering branch, × 1; **13**, hairs from corolla, × 30; **14**, corolla-lobe, × 3; **15**, stamen, × 3; **16**, style-tip, × 3; **17**, fruit, × 3. 1–8, from *Polhill* 4855; 9, 10, from *Tweedie* 1131; 11, from *Perdue & Kibuwa* 9018; 12–16, from *Bullock* 1932; 17, from *Polhill* 4854. Drawn by Christine Grey-Wilson.

TANZANIA. Lushoto, near Boma, 20 July 1962, *Semsei* 3485!; Morogoro District: Mkuyuni, 24 June 1983, *Polhill & Lovett* 4917!, Njombe District. Mdando Forest, between Madilu and Ilininda villages E. of Njombe–Manda road at Lusitu, 6 Oct. 1988, *Wiens & Calvin* 7005!; Zanzibar I., without precise locality or date, *Dowson* 601!

DISTR. **T** 3, 6, 7; **Z**; Malawi and N. Mozambique

HAB. Montane forest, extending down to coast on Mafia I., Zanzibar I. and in some low-lying mist-forests between Dar es Salaam and Morogoro, on many hosts, commonly parasitising plantation crops; 0–2400 m.

SYN. *Loranthus sigensis* Engl. in E.J. 20: 101 (1894) & P.O.A. C: 166 (1895); Sprague in F.T.A. 6(1): 289 (1910); F.D.O.-A. 2: 162 (1932); T.T.C.L.: 288 (1949)

    *L. wentzelianus* Engl. in E.J. 28: 381 (1900); Sprague in F.T.A. 6(1): 289 (1910); F.D.O.-A. 2: 163 (1932); T.T.C.L.: 288 (1949). Type: Tanzania, Morogoro District, Uluguru Mts., Lukwangule Plateau, *Goetze* 278 (B!, holo., K!, fragment)

    *L. berliniicola* Engl. in E.J. 28: 382 (1900); Sprague in F.T.A. 6(1): 290 (1910); F.D.O.-A. 2: 163 (1932); T.T.C.L.: 287 (1949). Type: Tanzania, Iringa District, Matanana Plateau, *Goetze* 742 (B!, holo., BR!, iso., K!, fragment)

    *Tapinanthus berliniicola* (Engl.) Danser in Verh. K. Akad. Wet., sect. 2, 29(6): 108 (1933)

    *T. sigensis* (Engl.) Danser in Verh. K. Akad. Wet., sect. 2, 29(6): 120 (1933)

    *T. wentzelianus* (Engl.) Danser in Verh. K. Akad. Wet., sect. 2, 29(6): 122 (1933)

NOTE (on species as a whole). Closely related to *P. regularis* and other Ethiopian species, but with shorter heads to corolla-buds and shorter corolla-lobes. The essential features of the flowers are very constant, but in other respects the species is polymorphic, with considerable variation in the shape of the leaves and development of indumentum. Eastern plants, from the Usambara Mts. south to the mountains around Lake Malawi, are recognised as a separate subspecies, characterised principally by the evanescent and appressed indumentum of the corolla, but flowers which are nearly as glabrous do occur sporadically along the Western Rift Valley and the Eastern Rift Valley in Kenya. The ranges of the two subspecies cross in the Southern Highlands of Tanzania. Subsp. *usuiensis* goes east from the Ufipa Plateau through Tukuyu and Songea to the Rondo Plateau near the coast. It is not recorded above 1800 m. in **T** 7. Subsp. *sigensis* comes south from the Nguru and Uluguru Mts., along the Iringa scarp, through Dabaga, Mufindi and Njombe to the Livingstone Mts. In **T** 7 it is recorded above 1800 m., though it does extend down to drier mist-forests in **T** 6, from Mahenge eastwards. The morphological differences seem to hold in these provinces, though some specimens have been misidentified and need to be relabelled. *P. cornetii* grows in the *Brachystegia* woodlands of southern **T** 7, but has distinctly longer hairs on the corolla.

3. **P. cornetii** (*Dewèvre*) *Polh. & Wiens* in Lebrun & Stork, Énum. Pl. Fl. Afr. Trop. 2: 170 (1992) & Mistletoes Afr.: 260, photo. 13, 114 (1998). Type: Zaire, Shaba, *Cornet in Descamps* (BR!, holo.)

Branches 1–2 m. long, the longer ones pendent; branchlets tomentose with red-brown dendritic hairs 0.5–1.5 mm. long, tardily glabrescent. Petiole 0.5–2 cm. long; lamina coriaceous, ovate to elliptic or elliptic-oblong, 8–20 cm. long, 3.5–9 cm. wide, obtuse to shortly rounded at the apex, cuneate to rounded or shortly cordate at the base, with fine felty tomentum when young of grey and red-brown stellate hairs, with some short dendritic hairs medially beneath, glabrescent, with 6–10 pairs of inconspicuous lateral nerves. Umbels confluent around older nodes mostly below current leaves, sessile or subsessile, 2–4-flowered; pedicels 0–1.5 mm. long; bract ovate to ovate-lanceolate, 3–6 mm. long, hairy. Receptacle 2–3.5 mm. long, red-brown tomentose; calyx (0.5–)1–1.5 mm. long, tomentose, ultimately sometimes somewhat glabrescent. Corolla 3.5–5 cm. long, coloured by red-brown tomentum outside, reddish on head of bud, pink to orange-red inside, red along filament-lines and darkening red overall, red-brown tomentose with dendritic hairs to 1.5 mm. long and sometimes subsimple hairs to 3 mm. long, slightly glabrescent on head of bud; apical swelling of bud ovoid to globose-ellipsoid, 3.5–5 mm. long, 3–3.5 mm. in diameter, pointed; basal swelling 6–7 mm. long, 4–5 mm. in diameter; lobes erect, 8–12 mm. long; tips elliptic, apiculate, 3–4 mm. long, 1.5–2 mm. wide, hardened inside. Filaments dark red; anthers 1.2–2 mm. long, with 5–7 cells in each row. Style abruptly expanded to twice its diameter opposite top of corolla-tube, then tapered slightly to the neck; neck 1.2–2 mm. long. Berry ellipsoid, 1.5 cm. long, 1 cm. in diameter, reddish tomentellous.

TANZANIA. Mbeya District: Pungaluma Hills, above Mshewe village, 18 Nov. 1989, *P. Lovett & Kayombo* 3340!; Rungwe District: Mulinda, 30 July 1912, *Stolz* 1477!
DISTR. **T** 7; Zaire (Shaba), Angola, Zambia and Malawi
HAB. *Brachystegia* woodland, generally parasitic on *Uapaca*; 900–1400 m.

SYN. *Loranthus cornetii* Dewèvre in B.S.B.B. 34(2): 92 (1895); Sprague in F.T.A. 6(1): 285 (1910); Balle in F.C.B. 1: 368, t. C, D facing 346 (1948)
    *L. usuiensis* Oliv. var. *longipilosus* Engl. & K. Krause in E.J. 51: 460 (1914); F.D.O.-A. 2: 162 (1932); T.T.C.L.: 288 (1949). Type: Tanzania, Rungwe District, Mulinda, *Stolz* 1477 (B!, holo., K!, WAG!, iso.)
    *Tapinanthus cornetii* (Dewèvre) Danser in Verh. K. Akad. Wet., sect. 2, 29(6): 110 (1933)

NOTE. Similar to *P. usuiensis*, which occurs in forests of the same region, but easily distinguished by the style enlarged in the upper part, also with broader corolla-lobes and altogether more rufous hairy.


4. **P. regularis** (*Schweinf.*) *M.G. Gilbert* in Nordic Journ. Bot. 5: 224 (1985); Balle in Stuttg. Beitr. Natur. 221: 3 (1970), *nom. invalid.*; M.G. Gilbert in Fl. Ethiopia 3: 368, fig. 114.8/4–5 (1990); Polh. & Wiens in U.K.W.F., ed. 2: 157 (1994) & Mistletoes Afr.: 262, photo. 117 (1998). Type: Ethiopia, Adowa, *Schimper* II. 747 (B!, holo., BM!, BR!, FT!, K!, MO!, P!, iso.)

Branches mostly pendent, 1–3 m. long; branchlets tomentose with yellow-brown dendritic hairs, glabrescent. Petiole stout, 1.5–3(–4) cm. long; lamina dark olive-green to grey-green, coriaceous, ovate to elliptic-oblong, 5–20(–23) cm. long, 3.5–11 cm. wide, rounded or slightly acuminate to a blunt tip, broadly cuneate to rounded or slightly cordate at the base, tawny tomentose when very young, becoming scurfy to varying degrees with minute grey stellate hairs, more persistently so beneath, with 6–10 pairs of inconspicuous lateral nerves. Umbels (1–)several confluent at old nodes, sessile, 4–8-flowered; pedicels 0–2 mm. long; bract ovate-elliptic, subacuminate at the apex, 4–6 mm. long, tomentellous. Receptacle broad, 2.5–3 mm. long, yellow-brown tomentellous, with longer dendritic hairs at the base or overall; calyx 1–1.2 mm. long, slightly glabrescent, long-ciliate. Corolla 4–5.8 cm. long, orange-yellow with brownish tomentum, orange to red on filament-lines inside, tomentose with grey stellate, tawny dendritic and subsimple hairs up to 2 mm. long, longer hairs most evident at the base and most evanescent at the apex; apical swelling of bud ellipsoid, ± apiculate, 5–6 mm. long, 3–3.5 mm. in diameter; basal swelling 5–6 mm. long, 4–4.5 mm. in diameter; lobes erect, 1.2–1.6 cm. long; tips narrowly elliptic, apiculate, 4.5–5 mm. long, 1.5–1.8 mm. wide, hardened inside. Filaments red below, green above; anthers 3–3.5 mm. long, with 5–6 cells in each row. Style slightly broadened upwards; neck 3–3.5 mm. long. Berry ellipsoid, 1.2–1.5 cm. long, 0.8–1 cm. in diameter, with bluish ground-colour and brown tomentellous, ultimately glabrescent; seed red.

KENYA. Laikipia, 1925, *Battiscombe* in F.D. 1126!; N. Nyeri District: 50 km. Nanyuki–Isiolo, 27 Mar. 1978, *M.G. Gilbert* 5038! & W. Mt. Kenya, Rongai R., 12 Jan. 1922, *R.E. & T.C.E. Fries* 869!
TANZANIA. Arusha District: Mt. Meru, Arusha National Park, Jckukumia, 6 Nov. 1969, *Richards* 24615! & Ngurdoto Crater gate, 13 Oct. 1965, *Greenway & Kanuri* 12132!; Moshi District: N. Kilimanjaro, Rongai, 18 Apr. 1934, *Schlieben* 5087!
DISTR. **K** 3, 4; **T** 2; Ethiopia, Yemen and Saudi Arabia
HAB. Upland dry evergreen forest and associated woodland, on various hosts including *Juniperus*; 1350–2500 m.

SYN. [*Loranthus rufescens* sensu A. Rich., Tent. Fl. Abyss. 1: 339 (1848), *non* DC.]
    *L. regularis* Schweinf. in Bull. Herb. Boiss. 4, append. 2: 148 (1896); Sprague in F.T.A. 6(1): 287 (1910); K. Krause in N.B.G.B. 8: 497 (1923), pro parte

*Tapinanthus regularis* (Schweinf.) Danser in Verh. K. Akad. Wet., sect. 2, 29(6): 118 (1933)

NOTE. Although *P. regularis* slightly overlaps the range of *P. usuiensis* in East Africa it occurs here only in the drier *Juniperus* forests on the rain-shadow side of the Rift Valley mountains.
In Mistletoes of Africa we cited the basionym authority as Steud. ex Sprague, following M.G. Gilbert in Fl. Ethiopia, but C. Whitehouse has kindly pointed out that Schweinfurth made a valid combination twenty years previously.

5. **P. dschallensis** (*Engl.*) *M.G. Gilbert* in Nordic Journ. Bot. 5: 224 (1985); Blundell, Wild Fl. E. Afr.: 130, t. 513 (1987); M.G. Gilbert in Fl. Ethiopia 3: 368 (1990); Polh. & Wiens in U.K.W.F., ed. 2: 157, t. 56 (1994) & Mistletoes Afr.: 261, photo. 115, fig. 27C (1998). Type: Tanzania, Moshi District, Lake Chala [Dschallasee], *Volkens* 1776 (B!, holo., BM!, G!, K!, iso.)

Branches mostly pendent, 0.7–2 m. long; branchlets tomentellous with short dendritic mostly red-brown hairs, soon falling to leave a covering of grey stellate hairs and scales, glabrescent. Leaves subopposite to irregularly ternate, subsequently clustered on short shoots; petiole 2–10 mm. long; lamina grey-green, coriaceous, linear-lanceolate to narrowly elliptic-oblong, 3–15 cm. long, 0.5–1.5(–2.5) cm. wide, tapered at either end, acute to obtuse at the apex, with minute grey stellate hairs and very short evanescent red-brown dendritic hairs, glabrescent, with 4–8 pairs of lateral nerves soon becoming obscure. Umbels confluent around old axils, mostly after primary leaves have fallen and often below tufts of new leaves, sessile, 2–4-flowered; pedicels 1–3 mm. long; bract ovate, ± acuminate at the apex, 3–5 mm. long, tomentellous to glabrate. Receptacle obconic to broadly barrel-shaped, 2.5–3 mm. long, tomentellous; calyx 1.5–2 mm. long, truncate to slightly lobed, less hairy, ciliate. Corolla 4.5–6 cm. long, orange-red with red filament-lines, lobes sometimes reddish but tips cream inside, tomentellous with grey stellate hairs and very short red-brown dendritic hairs less than 0.5 mm. long, often slightly glabrescent on basal swelling; apical swelling of bud ovoid-ellipsoid, apiculate, 4–5 mm. long, 3.5–4 mm. in diameter, slightly 5-ribbed; basal swelling 5–7 mm. long, 4 mm. in diameter; lobes erect, 1.1–1.3 cm. long; tips elliptic, 4–5 mm. long, 2–2.5 mm. wide, conspicuously hardened inside. Filaments red on lower part; anthers 2–2.8 mm. long, with 6–8 cells in each row. Style red, generally distinctly lobed opposite top of corolla-tube, slightly tapered to neck; neck 2–2.8 mm. long. Berry blue-green, ellipsoid, 1.5 cm. long, 1 cm. in diameter, with fine rufous stellate pubescence; seed blue-black. Fig. 18/12–17 (p. 107).

UGANDA. Karamoja District: Moroto, 2 Jan. 1937, *A.S. Thomas* 2139! & Kakamari, June 1930, *Liebenberg* 384!; Mbale District: Bugishu, Greek River Camp, Jan. 1936, *Eggeling* 2507!
KENYA. W. Suk District: 12 km. N. of Makutano, 26 Dec. 1971, *Wiens* 4504!; N. Nyeri District: 9 km. from Timau on old Isiolo road, 21 Apr. 1981, *M.G. Gilbert* 6005!; Kericho District: 3 km. S. of Chemelil, 14 Mar. 1985, *Kirkup* 37!
TANZANIA. Musoma District: Serengeti National Park, Duma R., Mamerehe Guard Post, 11 Nov. 1961, *Greenway* 10253!; Kondoa District: Great North Road, 8 km. S. of Bereku, 9 Jan. 1962, *Polhill & Paulo* 1112!; Mbeya District: Pungaluma Hills, above Mshewe village, 10 Dec. 1989, *Lovett, Sidwell & Kayombo* 3643!
DISTR. U 1, 3; K 2–6; T 1, 2, 4, 5, 7; Somalia and Zambia
HAB. Woodland, bushland and wooded grassland, generally on *Acacia*; 750–1950 m.

SYN. *Loranthus sp. nov.?* sensu Oliv. in Trans. Linn. Soc., Bot., 29: 81 (1873)
*L. dschallensis* Engl., P.O.A. C: 166 (1895); Sprague in F.T.A. 6(1): 286 (1910); F.D.O.-A. 2: 162 (1932); T.T.C.L.: 287 (1949); U.K.W.F.: 330, fig. on 329 (1974)
*Tapinanthus dschallensis* (Engl.) Danser in Verh. K. Akad. Wet., sect. 2, 29(6): 111 (1933)

6. **P. polycrypta** (*Didr.*) *Balle* in K.B. 11: 168 (1956) & in F.W.T.A., ed. 2, 1: 664 (1958), pro parte, & in Fl. Cameroun 23: 40, t. 10/1–6 (1982), excl. subsp. *raynaliana* & var. *luteovittata*; Polh. & Wiens, Mistletoes Afr.: 264 (1998). Type: Zaire, *C. Smith* (C, holo., K, fragment & drawing!)

Branches mostly pendent, 1–2 m. long; branchlets tomentellous with rusty brown dendritic hairs less than 0.5 mm. long. Petiole 5–20 mm. long; lamina coriaceous, discolorous, upper side sometimes dark bluish green, lanceolate, oblong-lanceolate, elliptic-oblong or ovate, 4–16 cm. long, 1.5–4 cm. wide, acutely to bluntly pointed at the apex, rounded to cordate at the base, soon glabrescent and glossy above, persistently tomentellous beneath with pale stellate hairs and short red-brown dendritic hairs, with 8–12 pairs of generally obscure lateral nerves. Umbels mostly crowded at old nodes, normally 2–4-flowered; peduncle 0–5 mm. long; pedicels 0–4 mm. long; bract lanceolate to ovate, 2–4 mm. long. Receptacle 1.5–2 mm. long, tomentellous; calyx 0.5–1 mm. long, sometimes slightly toothed, generally tomentellous. Corolla 3 1.2 cm. long, yellow, generally orange or red on the tube and tip of the lobes, covered with small stellate to shortly dendritic hairs, glabrescent on apical and basal swellings or subglabrous overall; apical swelling of bud obovoid-ellipsoid to ovoid, ± 3 mm. long, 2.5 mm. in diameter, 5-ribbed; basal swelling 4–5 mm. long, 2.5 mm. in diameter; lobes erect, 9–12 mm. long, with relatively short spathulate tips hardened inside. Anthers 1–1.5(–1.8) mm. long, with 3–4(–6) cells in each row. Style gradually broadened opposite the filaments, generally densely papillate; neck 1–1.5(–1.8) mm. long; stigma ovoid to depressed conical, generally at least as broad as shoulders. Berry ellipsoid-globose to obovoid, 8–10 mm. long, 6–7 mm. in diameter, russet-pink, subpersistently tomentellous.

SYN. *Loranthus polycryptus* Didr. in Vidensk. Medd. Dansk. Naturhist. Foren. Kjöbenh. 1854: 194 (1854); Sprague in F.T.A. 6(1): 284 (1910); Balle in F.C.B. 1: 360 (1948)
   *L. discolor* Engl. in B.S.B.B. 39: 26 (1900); Sprague in F.T.A. 6(1): 284 (1910); Balle in F.C.B. 1: 361 (1948), *nom. illegit.*, *non* Schinz (1896). Type: Zaire, Pool de Maleba [Stanley Pool], *E. Laurent* (BR!, holo., B!, iso., K!, fragment)
   *Tapinanthus polycryptus* (Didr.) Danser in Verh. K. Akad. Wet., sect. 2, 29(6): 117 (1933)
   *T. pseudonymus* Danser in Verh. K. Akad. Wet., sect. 2, 29(6): 118 (1933). Type as for *Loranthus discolor* Engl., *non Tapinanthus discolor* (Schinz) Danser
   *Loranthus polycryptus* Didr. var. *discolor* Balle in B.J.B.B. 17: 229 (1944). Type as for *L. discolor* Engl.

subsp. **subglabriflora** *Polh. & Wiens*, Mistletoes Afr.: 265, fig. 27E (1998). Type: Uganda, Mengo District, 22 km. Kampala–Entebbe, *Chandler* 2095 (K!, holo., B!, BM!, P!, iso.)

Corolla subglabrous.

UGANDA. Kigezi District: Kayonza, May 1950, *Purseglove* 3451!; Mbale District: Sipi, 12 Dec. 1938, *A.S. Thomas* 2604!; Mengo District: Entebbe, Sept. 1922, *Maitland* 640!
TANZANIA. Bukoba District: Ruzinga, Kikuru Forest, July 1991, *Kielland* 18! & Rusinga Forest Reserve, Kagera, 23 Dec. 1994, *Congdon* 391!
DISTR. U 2–4; T 1; Gabon, Zaire and Sudan
HAB. Forest, sometimes riverine or in secondary associations nearby; 900–1800 m.

NOTE. Subsp. *polycrypta*, with a hairy corolla, occurs W. of the Ubangui R. in Cameroon, Gabon, Congo and the Central African Republic, then through western and southern Zaire to Angola and western Zambia. The ranges overlap in Gabon, where subsp. *subglabriflora* extends to the coast.

7. **P. eminii** *(Engl.) Polh. & Wiens* in Lebrun & Stork, Énum. Pl. Fl. Afr. Trop. 2: 171 (1992); Mistletoes Afr.: 271, photo. 121A–B, fig. 28C (1998). Type: Tanzania, Mwanza District, Bukumbi, *Stuhlmann* 819 (B!, holo., K!, P!, fragments)

Stems in all directions to 60 cm., much branched; branchlets tomentose, with buff dendritic hairs 0.5–1 mm. long on youngest parts, and shorter hairs that are paler and more persistent. Petiole 5–10 mm. long; lamina ovate-lanceolate to ovate-elliptic to ovate, 2–7 cm. long, 1.5–4 cm. wide, subacute to obtuse or shortly rounded at the apex, cuneate to rounded or slightly cordate at the base, tomentose with stellate and shortly dendritic hairs, glabrescent to varying degrees above, persistently hairy beneath, often with slightly longer and more tawny hairs especially medially, with 4–6 pairs of obscure lateral nerves. Flowers progressively produced in axils; pedicels 2–4 mm. long; bracts ovate-lanceolate to shortly spathulate, 3–6 mm. long, occasionally some foliaceous. Receptacle 2–4 mm. long, tomentose to villous with pale-coloured to violet hairs; calyx 1–2 mm. long, subentire to slightly lobed, often somewhat glabrescent, ciliate. Corolla 4–5 cm. long, pale green, generally red at least on the claws of the lobes inside, with an understorey of pale hairs and many longer pale, reddish or purplish dendritic to subsimple hairs 1–4 mm. long; apical swelling of bud ellipsoid, 5–7 mm. long, 3–3.5 mm. in diameter; basal swelling slight; lobes reflexed, 9–12 mm. long, the upper half to three-fifths linear-lanceolate, hardened inside. Filaments generally pale green; tooth 0.5–0.8 mm. long; anthers 2–3 mm. long, with 6–8 cells in each row. Style pale green, slender, only slightly broadened opposite the filaments, minutely to distinctly denticulate-papillate, narrowed 2–3.5 mm. below the obovoid stigma. Berry ellipsoid, with persistent calyx, 7–9 mm. long, 4–5 mm. in diameter, shortly hairy, turning blue-green. Fig. 19/1–7.

TANZANIA. Old Shinyanga, 14 Apr. 1953, *Welch* 1965!; Mpanda District: Sitebi Hill, 10 Feb. 1996, *Congdon* 453a!; Kondoa District: Kolo, 12 Jan. 1962, *Polhill & Paulo* 1153!
DISTR. T 1, 4, 5, 7; Zaire (Shaba), Zambia, Malawi, Mozambique and E. Zimbabwe
HAB. Plateau woodland and riverine associations, on various hosts; 1150–1900 m.

SYN. *Loranthus eminii* Engl. in E.J. 20: 113 (1894) & P.O.A. C: 166, t. 16D–K (1895); Sprague in F.T.A. 6(1): 332 (1910); F.D.O.-A. 2: 170 (1932); Balle in F.C.B. 1: 368 (1948); T.T.C.L.: 287 (1949)
   *Tapinanthus eminii* (Engl.) Danser in Verh. K. Akad. Wet., sect. 2, 29(6): 111 (1933)
   *Loranthus eminii* Engl. forma *cinereus* Balle in B.J.B.B. 17: 236 (1944). Type: Zaire, Shaba, Lubumbashi [Elisabethville], *Quarré* 6195 (BR!, holo.)
   *L. eminii* Engl. forma *rufescens* Balle in B.J.B.B. 17: 236 (1944). Type: Zaire, Shaba, Kapiri valley, *Homblé* 1244 (BR!, holo.)

NOTE. A distinctive but variable species, showing incipient divergence throughout its range, most evident in differences in indumentum. The typical form from south of Lake Victoria and at lower altitudes west of Lake Tanganyika has ovate-lanceolate leaves, rounded to slightly cordate at the base, with a short persistent indumentum, with clearly dendritic brownish hairs on the corolla, in NW. Tanzania becoming more wispy, less regularly branched and paler further south-east in Tanzania to Zambia. On higher ground around the southern end of Lake Tanganyika and sporadically southwards to Zimbabwe and west to Shaba there is a form with more variably shaped leaves tending to be more glabrescent above, with rather long pale hairs on the corolla, and variably tomentose to villous receptacles. An extreme form around Lake Tanganyika at higher altitudes is essentially the same, but the hairs are violet-purple, often long on the receptacle as well as on the corolla.

8. **P. proteicola** (*Engl.*) *Polh. & Wiens* in Lebrun & Stork, Énum. Pl. Fl. Afr. Trop. 2: 171 (1992) & Mistletoes Afr.: 272, photo. 122, fig. 28D (1998). Type: Tanzania, Njombe District, Lipanga Mt., *Goetze* 995 (B!, holo., BR!, P!, iso., K!, fragment)

Stems in all directions to 1 m. or so, much branched; branchlets soon stout, tomentose, with longer orange-tawny dendritic to subsimple hairs 1–4 mm. long on youngest parts. Leaves and leafy bracts rather crowded towards ends of branches; petiole 0.5–1.5 cm. long; lamina softly coriaceous, lanceolate to ovate-oblong or elliptic, 3–10 cm. long, 1.5–3 cm. wide, obtuse to subacute at the apex, cuneate to slightly rounded at the base, covered with inconspicuous greyish stellate and shortly dendritic hairs above, pale tomentellous beneath with slightly longer more tawny hairs at least medially, with 4–8 pairs of obscure lateral nerves. Flowers 2–several in the axils, subsessile; pedicels mostly 0.5–1 mm. long; bracts mostly spathulate to foliaceous, 1–1.5 cm. long, tomentellous. Receptacle 2.5–3 mm. long, densely villous; calyx 1.5–2 mm. long, similarly hairy but somewhat glabrescent, long-ciliate. Corolla

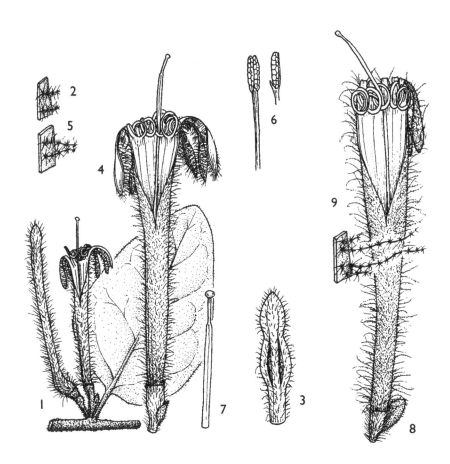

FIG. 19. *PHRAGMANTHERA EMINII* — **1**, flowering node, × 1; **2**, hairs from leaf, × 20; **3**, tip of mature flower-bud, × 4; **4**, flower, × 2; **5**, hairs from corolla, × 6; **6**, anther, two views, × 4; **7**, style-tip, × 4. *PHRAGMANTHERA PROTEICOLA* — **8**, flower, × 2; **9**, hairs from corolla, × 6. 1–7, from *Brummitt et al.* 15969; 8, 9, from *Milne-Redhead & Taylor* 10810. Drawn by Christine Grey-Wilson.

4.5–6 cm. long, usually red at least on the claw of the lobes inside, otherwise cream to greenish, tomentellous with pale dendritic hairs and also sparingly villous with tawny-orange subsimple hairs 3–4 mm. long, also with a tuft of long hairs at the base; apical swelling of bud slight, 8–9 mm. long, 3 mm. in diameter; basal swelling slight; lobes reflexed, 10–12 mm. long, the upper half linear-elliptic, pointed, hardened inside. Filaments green to red; tooth 0.7–1 mm. long; anthers 3–4 mm. long, with 6–12 cells in each row. Style pale green, slender, slightly thickened opposite the filaments, conspicuously glandular-papillate, narrowed to a 3–4 mm. long neck below the ovoid to obovoid stigma. Berry ellipsoid to obovoid, 8 mm. long, 5 mm. in diameter, villous. Fig. 19/8, 9.

TANZANIA. Ufipa District: Wipanga, 14 km. NNE. of Sumbawanga, 16 June 1987, *Mwasumbi, Mohamed & Kajula* 13117! & Mbaa Mt., 23 Nov. 1995, *Congdon* 444!; Songea District: Matengo Hills, Miyau, 3 Mar. 1956, *Milne-Redhead & Taylor* 8952!
DISTR. **T** 4, 7, 8; Zambia and Malawi

Hab. Generally in higher rainfall *Brachystegia* woodland, often on scarps and rocky hills, commonly on *Ochna*, *Faurea* and *Protea*, but descending along rivers in places; 900–2150 m.

Syn. *Loranthus proteicola* Engl. in E.J. 30: 303 (1901); Sprague in F.T.A. 6(1): 333 (1910); F.D.O.-A. 2: 170 (1932); T.T.C.L.: 287 (1949)

    *Tapinanthus proteicola* (Engl.) Danser in Verh. K. Akad. Wet., sect. 2, 29(6): 118 (1933)

Note. Closely related to the more variable *P. eminii*, but generally easily recognised by the subsessile flowers, leafy bracts, longer anthers, more papillate style, more crowded and often rather elongate leaves. The record mapped in Polh. & Wiens (1998) for **T** 5 was based on a misidentified specimen of *P. eminii* (*Richards* 1991), which admittedly does not have the characters perfectly correlated, but matches better with other collections from the same province.

# INDEX TO LORANTHACEAE

**No new names validated in this part**

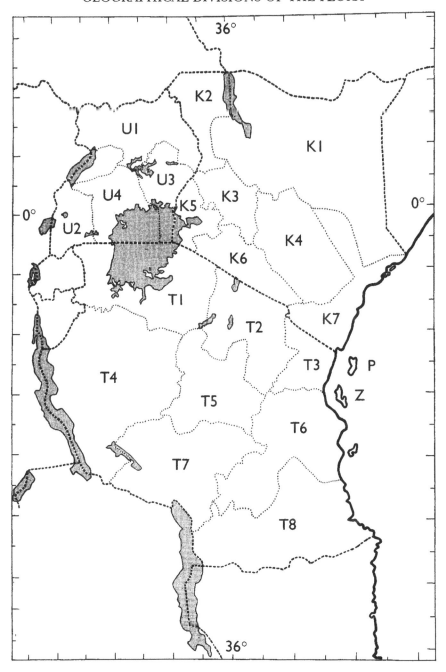

T - #0603 - 101024 - C0 - 234/156/7 - PB - 9789061913832 - Gloss Lamination